★ ★ ★ # STABLE SMARTS ★ ★ ★

Stable Smarts

★ ★ ★ ★ ★

Sensible Advice, Quick Fixes,
and Time-tested Wisdom
from an Idaho Horsewoman

★ ★ ★ ★ ★

HEATHER SMITH THOMAS

Illustrations by Gregory Wenzel

Storey Publishing

The mission of Storey Publishing is to serve our customers
by publishing practical information that encourages personal independence
in harmony with the environment.

Edited by Cindy A. Littlefield
Art direction by Vicky Vaughn and Kent Lew
Cover design by Carol Jessop, Black Trout Design
Text design by Carol Jessop and Jessica Armstrong
Text production by Jessica Armstrong
Illustrations by © Gregory Wenzel
Additional graphic elements by © Carol Jessop
Indexed by Terri Corry, Stepping Stone Indexing Services

Printed in the United States by Von Hoffmann
10 9 8 7 6 5 4 3 2 1

Library of Congress Cataloging-in-Publication Data

Thomas, Heather Smith, 1944–
   Stable smarts / by Heather Smith Thomas ; illustrations by Gregory Wenzel.
      p. cm.
   Includes bibliographical references and index.
   ISBN-13: 978-1-58017-610-1; ISBN-10: 1-58017-610-0 (pbk. : alk. paper)
   1. Horses. 2. Horses—Miscellanea. I. Title.

SF285.3.T493 2005
636.1—dc22

                                          2005021728

Contents

★　　★　　★

This book is dedicated to my husband, Lynn Thomas — my help-
mate for 40 years of raising cattle and horses. Together we made
do or came up with ideas to make things work, whether it was patching a fence, repairing horse gear, creating a special shoe for a horse
with a foot problem, or taking care of a sick or injured animal that
presented a unique challenge. Not all the tips and handy hints in this
book originated with us; we're always open to good ideas and suggestions passed along by other horsemen. But Lynn has been my "fix it"
man whenever I needed to make something work or find a way around
a problem, and his innovations have often saved the day.

When working with horses, there is always more to learn. This book
is a gathering together of great ideas, handy tips, good-sense advice,
and horsekeeping lore that will be helpful to all horsemen. Both
novice and professional will find useful information here — ideas and
tips gleaned from a lifetime of working with horses and from the inno-
vative suggestions of others.

Many of these handy ideas and bits of advice are not found in publi-
cations on horse care. I hope that this book will be a treasure trove
and handy reference for many time/energy/money-saving ideas, safety
tips, commonsense shortcuts, problem-solving suggestions, and other
advice that can make your life with horses much easier. The various
chapters put these handy hints in an organized and easy-to-find form
and deal with many aspects of horse care and handling that the horseman will encounter.

★　　★　　★

CHAPTER ONE

Feed & Water

There are many factors involved in making sure
a horse gets all the nutrients he needs to stay fit and
healthy. Here are tips for deciding what, how,
and when to feed him.

HAY

Hay is the mainstay of a horse's diet when he is not on pasture. Therefore it's important that the hay you feed him be of good quality to provide all the essential nutrients he needs.

★　　★　　★

Stacking Outdoors

When stacking hay in the open rather than in a barn or shed, protect it from spoiling by placing the bottom bales on wood pallets or poles laid on the ground. Shipping pallets, which many business routinely discard, can often be picked up at no charge. For an additional moisture barrier, cover the pallets with a tarp and stack the hay on it.

Never place bales directly on the ground for more than a day unless they're on a layer of gravel, which will provide drainage. Otherwise, the hay will draw moisture from the ground, and rainwater may seep underneath. If hay is stacked on grass or any other moist ground cover, the bottom of the bales will mold quickly.

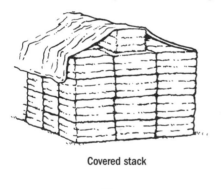

Covered stack

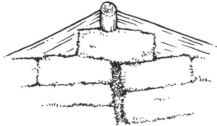

Peaked tarp roof

Protect the top of the stack by covering it with a tarp properly tied down so it won't blow off. First make a sloped roof: Create a peak in the center either by stacking bales higher there or by setting ridgepoles on top of thick flakes of straw or hay. Before placing the tarp over the stack, tie baling twine to each grommet so that once the "roof" is in place you can fasten it securely to the sides of the hay-stack. Use a ladder to take the ridge supplies to the top of the stack, keeping in mind that straw is lighter than hay and easier to carry. When you stretch the tarp tightly over the ridge, the tentlike roof will shed water, and there'll be no place for the moisture to puddle. Even if there are some holes in the tarp, most of the water will run off, and the airspace between the tarp and the hay will allow moisture to evaporate.

Storing in a Barn

It's safest to store hay in a separate building or hay shed, away from the horses. When hay is stored in the same barn as the horses, it should be in a walled-off area constructed of cinder or concrete blocks, with a fireproof wall between it and the rest of the barn, and steel fire doors that are always kept closed when not in use.

If you store hay in a loft, it can be a serious fire hazard unless the loft has a fire-resistant floor, such as sheets of drywall sandwiched between plywood.

Preventing Hay Fires

Every year many haystacks and barns are lost to fire caused by spontaneous combustion, a chemical process that occurs when damp hay ferments, heats, and ignites. Frequent rains may make it difficult to dry hay enough to store it safely. In fact, the threat of rain itself may cause a farmer to bale and stack hay that's still too green. Grass hay can usually be baled slightly greener than legume hay, which tends to mold and heat more readily because it's richer in nutrients.

ABSORBING MOISTURE WITH SALT

One way to help prevent mold is to sprinkle rock salt over a tarp or on the floor of your storage shed or hay loft before you place the hay on it. If you live in a damp climate, also try sprinkling rock salt (about 1 cup per six bales) between each layer of bales to help absorb moisture.

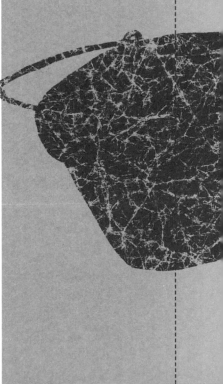

The larger the bale or the stack, the greater the danger of the hay heating and igniting, because less of it is exposed to the air, where heat can escape. Large bales (900 pounds or more) have been known to heat enough to ignite while still standing in the field before stacking. And the tighter the bale, the greater the risk of heating: Any wet material sealed into an airtight package will hasten the combustion process. When you're buying big bales, be sure they were harvested under optimum conditions and are fully dry, because each is like a small, compact haystack.

But any size bale is subject to heating when it's too green or when rain has penetrated several inches into it in the field. Such bales should not be stacked until they're completely dry, or they may continue heating. Always check the interior, and if you detect any heat, don't buy or store the bales in your barn.

Checking Bale Temperature

The higher the moisture level in hay, the hotter it gets and the longer it takes to cool. Hay fires may occur six weeks or longer after baling. In a small bale, the temperature of "hot" hay will usually peak at about 125°F within three to seven days, but in big bales or a large, tight stack the temperature may build up over a longer time and become hotter.

With a small bale, you can easily feel between the flakes with your hand if you suspect the hay is heating. With a large bale, put a candy thermometer into a length of metal pipe and insert it into the bale, checking it 10 or 15 minutes later. If the temperature is over 140°F, check it in a few hours to see if it's gone up or down. The temperature will usually drop back to safe levels (below 140°F) within 15 to 60 days, depending on the size and density of the bale or stack, as well as on air temperature and humidity. If it goes up to 180°F, combustion has already begun or is about to begin.

Storing Long Term

When hay prices are high, buying year-old hay (if available) for a cheaper price may be tempting, but not always wise. Horses fed old hay may not receive enough vitamin A or protein and may need green grass or supplements. Always buy hay that was harvested under good conditions, with no bleaching (and consequent loss of nutrients) or mold.

If your supply must last until the next haying season, select hay with high carotene (vitamin A) levels, or plan to supplement it. Keep the hay dry and out of the sunshine, and stack it so the oldest batch is always used first. Divide your storage area so that when one section is depleted you can stack new hay there while you use hay from another area.

Hauling with a Four-Wheeler Trailer

An easy way to haul hay or straw bales around your property is with a small trailer pulled by a four-wheeler. You can construct a trailer using the front axle from a small car, or the rear axle from a small light truck. Weld scrap metal into a triangular frame to hook on top of the axle, or have a local welder do it for you.

For the floor, use exterior plywood and bolt angle iron from front to back and across the bottom for support. Facing up, the angle iron along the sides will keep the bales from sliding off. A piece of angle iron pointing down across the back will protect the rear edge of the plywood. If you want, add another piece of plywood across the front and brace it with flat strap iron. The bed is the perfect size for hauling two bales side by side, or a stack of four to six bales.

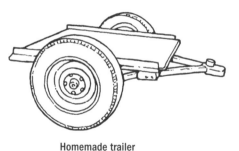

Homemade trailer

Most four-wheelers can pull 800 to 900 pounds, so you want the trailer itself to be light. Using small tires will reduce the trailer's weight as well as keep it low to the ground and make it easy to load and unload.

Hauling Bales in Winter

Pushing a wheelbarrow or wheeled cart full of hay through the snow is difficult. A child's sled works nicely for moving a single bale or two stacked bales if the ground is fairly level. The best kind of sled for moving barn cleanings is a molded plastic saucer.

Retying a Bale

To retie a partially used or a broken bale, fasten together the two strings at one end of the bale, using a square knot (see page 238). Run the two loose strings from the other end of the bale through the loop you cre-

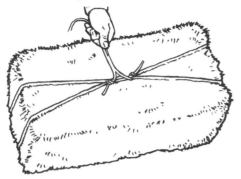

ated and pull the bale tightly together, then tie each string back to itself in a slipknot.

To secure a whole hay bale, either splice spare strings into the existing strings to lengthen them or split the bale into two smaller ones and retie.

Hay Hook Twine Cutter

Handling several bales at feeding time or loading hay onto a truck or wheelbarrow is easier when you use a hook. If you weld a tooth from a mowing machine blade to the back of the hook, it can double as a

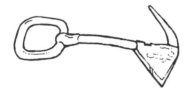

twine cutter. The mower tooth will stay sharp for years and eliminate the need to fumble in your pocket for a knife, a real benefit in winter when you're wearing gloves. If you don't weld, a machinery repair shop should be able to handle it for you quickly and simply.

ALFALFA SNAP TEST

A higher leaf-to-stem ratio in alfalfa hay means it's easier to digest, tastes better, and contains more nutrients. To check the quality of alfalfa hay, bend a handful of it. If the stems bend very easily, the fiber content is relatively low and the hay will be more nutritious. As the alfalfa matures, it loses nutrients and its stems become woodier, snapping like twigs.

Breaking Baling Wire

If you use hay bales bound by wire rather than twine, you can easily break the wire by slipping a hay hook under it in the middle of the bale and twisting it a couple of times. If the small twisted piece of wire breaks off, make sure it doesn't end up in the hay.

Removing Bale Strings without a Knife

To open a bale of hay when you don't have a pocket knife or twine cutter with you, try working a string off one corner. Once the twine is off that corner, it should be loose enough to remove from the other corner, and then the bale should be easy to break open.

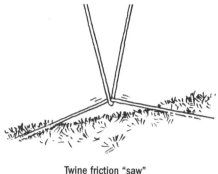

Twine friction "saw"

With a really tight bale, push a piece of baling twine under one of the bale strings to use as a friction "saw." Pulling the twine rapidly back and forth in the same spot will quickly cut the bale string in two.

Nitrate Poisoning in Cereal Grain

Cereal grain crops, especially oats, are sometimes cut for hay while they're still green and growing, rather than at maturity when they're typically harvested as grain. Oats cut before maturity can make very good hay. It's important, however, that oat hay is put up in ideal weather conditions. If the hay is damp, it will become moldy, especially if a legume, such as alfalfa or peas, is included in the mix to provide the protein that oats lack.

Even without legumes, cereal hay can be risky under certain conditions. There is always a chance of nitrate poisoning if the hay was cut after a growth spurt following a drought. Have the hay tested for nitrate content before you buy it; ask your Extension agent how to take a hay sample and where to send it.

TIPS *for*
SELECTING GOOD HAY

Keep these guidelines in mind when choosing hay.

★ The outside may be bleached, but it's what's inside that counts. Always break open a few bales to check. The interior should be green (not brown or yellow) and leafy (not coarse and stemmy). It should smell good and be soft to the touch.

★ Don't buy hay that is excessively dry and sun-bleached or baled too wet. If it smells musty, moldy, dusty, sour, or fermented, or has flakes that are brown or stuck together, don't feed it.

★ For the best quality, select hay that was cut before the grass made seed heads or the alfalfa blooms.

★ Avoid hay that contains weeds, dirt, or foreign material (such as sticks, baling twines, rocks, and wire).

★ Check for signs of insect infestation. Because blister beetles feed on alfalfa blossoms, avoid them by choosing hay that has not yet bloomed.

Meeting Your Horse's Nutritional Needs

Different horses have different nutritional needs. Young growing horses and lactating mares need more nutrients and protein than inactive, adult horses, which do better on more mature hay containing less protein and more fiber. Very fine-stemmed, leafy alfalfa (rabbit hay or dairy hay) is too rich and palatable for horses (they usually overeat it), and it doesn't have enough fiber for proper digestion. This type of alfalfa works best used sparingly as a supplement, for example, added to the diet of a young orphan foal or an older horse with poor teeth that can't eat hay very well.

Dangers of Weedy Hay

Some weeds are poisonous to horses and may cause indigestion and colic, or even death. Even if a horse avoids a toxic plant in the pasture, he may end up eating it once it's cut and dried and the offensive odor has diminished. If you find weeds in a bale, discard the portion containing them. Remember, not all horses are fussy enough to sort them out of hay, especially when they're hungry and it's the only food available.

Certain grasses are undesirable in hay. Downy brome (also called June grass or cheatgrass) and some foxtails, for instance, have sharp awns and seed heads that poke into the mouth tissues and cause sores or infections. If you inadvertently end up with some in your hay, discard all the flakes containing them.

Feeding Hay by Weight, not Volume

The average horse needs 1½ to 2 pounds of hay per 100 pounds of body weight, depending on the type and quality of hay. Therefore, a 1,000-pound horse needs 15 to 20 pounds of hay per day, split into several feedings. He needs less alfalfa than grass, because the alfalfa is more nutrient dense. Unless you adjust the flake size according to the type of hay being fed, you may be seriously overfeeding or underfeeding your horse. Also keep in mind that even though it takes a lot less alfalfa hay to provide good nutrition, your horse will still feel hungry because of the lack of roughage.

Here's an easy method for weighing hay bales: Weigh your pickup on a truck scale twice, first unloaded and then loaded with hay; divide the difference by the number of bales in the load. To weigh a flake of hay, stand on your bathroom scale holding the hay, and then again without it.

Homemade Corner Feeder

If your horse wastes hay when fed on the stall floor, try making a safe and inexpensive plywood corner feeder. Choose a corner away from the horse's waterer or bucket holder, so there'll be less chance of the hay getting wet.

To center the plywood, mark a spot on each wall about 33½ inches from the corner, then measure from one mark straight across to the other to find out how long the plywood should be. The height will depend on the size of the horse using the stall. You want it high enough so he won't step into the feeder, yet low enough so he can easily put his head in to eat. For the average horse, somewhere between 2 and 2½ feet high is about right.

Once it's in proper position, fitting nicely against the walls and floor, secure the plywood in place with wood screws or drywall screws. Next brace the feeder along the top edge with a 2 X 4 cut to fit the exact length of the plywood's top edge (taking into account the 45-degree angles created by the walls). Make sure the brace is level with the top of the plywood, and screw it to the wall and the plywood (from the plywood side) about every 8 to 10 inches. Use screws that are long enough to go deep into the brace but not come out the front where they can poke your horse. If he likes to crib or chew wood, secure a piece of angle iron to the front edge of the brace to protect it. Screw holes can be drilled through the angle iron with a metal drill bit (a mechanic, carpenter, or welder can help you with this if you don't have the proper tool).

Feeding a Group

A hay rack designed for horses to eat from the front and back, or from all sides, works well in a pen or pasture. A timid horse that's chased away by a dominant one can go around and eat on the other side. Provide more than one rack when there are more than three or four horses in the group. If you have only one feeder, put some hay on the ground in a clean area a distance away, to give the timid ones something to eat.

If the hay is not protected from weather, feed only the amount that will be eaten each day with none left to spoil. Or, feed extra; if horses always have a good supply, they can leave the less desirable hay. Remove any wasted hay regularly, and, if the area around the feeder becomes too muddy, move it to a new location.

If you feed on the ground, do so in clean, well-sodded areas. Spread the flakes to give the horses plenty of room to eat and to discourage one horse from chasing the others away. Always provide extra flakes for a timid horse to find.

Using Large Bales

Some horsemen use big round bales or ½-ton (or larger) square bales, but this method wastes a lot of hay if the bales are just set out in the pasture. You can expect about 40 percent to be scattered and wasted. When fresh hay is still available, most horses won't eat hay that's been tramped on.

Reduce the waste by putting a big bale in a feeder, though it may still become moldy or dusty if it gets wet. If you locate the feeder under a roof or in a round, open shed, the horses will be able to stand out of the rain (or in the shade) to eat. When the feeder is in a regular run-in shed and the group is large, there may be a dominance problem because not all of the horses will fit around the feeder.

Even with a feeder, there's bound to be some waste. For the best results, use a feeder designed for horses. The big, round bale feeders for cattle aren't as safe, because a horse can rub off his mane when reaching through the opening and injure his head when pulling out. A horse feeder, on the other hand, has loops for the horse to reach between rather than openings to put his head through. Check a feeder often to make sure it's in good repair, with no broken pieces or sharp areas.

Dealing with Dusty Hay

Many horses are allergic to mold spores in dusty hay and will develop a cough when it's fed to them. Alfalfa hay tends to be dustier than grass hay just because it's richer and more prone to mold. Even if you can't see any mold, a few spores may be released in the air when the horse shakes the hay around. Once a horse becomes sensitive to hay dust, even a single feeding may cause coughing and wheezing.

To avoid this problem, always select hay that is as dust free as possible. Shake it thoroughly, then settle any remaining dust particles with a fine spray of water before you feed your horse. If you don't have access to a hose, use a sprinkler can or improvise one by using a nail to punch a bunch of holes in the bottom of a coffee can; dip the can in a nearby ditch, stream, water bucket, or water trough.

Shaking the hay and sprinkling it with water are enough to defend most horses against respiratory problems, but a horse that already has an allergic sensitivity (heaves) may need more protection. Feed him only the best-quality grass hay and no alfalfa. Thoroughly dunk each flake in a tub or bucket of water, soaking the hay for several minutes or even a half hour, if necessary. An easy way to soak a horse's ration of hay is to put it in a nylon net and dunk it in a clean barrel or garbage can filled with enough water to completely immerse it. Change the water in the barrel or garbage can for each soaking. A pulley attached to the barn ceiling or an overhead pole rigged near your horse's pen will make it easy to lift the hay in and out of the container. Drain the hay briefly so it won't make a big puddle in the feeding area, and remove any uneaten portions of wet hay after the horse finishes.

Preventing Sand Colic

If the stall floor or pen is sandy, put a large rubber mat in front of the manger or feeder to keep a horse from eating sand with his last wisps of hay. Periodically sweep off the mat to keep it clean.

In a paddock without a feeder and no sod to set the hay on, put down a durable tarp to hold the hay. You may have to secure the edges and corners with dirt so the tarp won't blow away or wrinkle between feedings. Sweep or shake the tarp if it gets dirty or sandy.

Another option is to put the hay in a large cardboard box. The hay will stay clean, less hay will be tramped into the ground, and chewing on the cardboard won't hurt the horse. A box makes a handy feeder when you are traveling with a horse and feeding him while he's tied to the trailer. It'll easily fit in your tack compartment, and you can store other supplies in it when you're not using it as a feeder.

Lawn Clipping Safety

Lawn clippings usually aren't a safe feed for horses. With a moisture content of about 80 percent, they spoil quickly, fermenting within three or four hours when they're piled before they dry. The fermentation process may also produce nitrates that can be deadly, especially in clippings from a lawn that has been fertilized.

The safest way to feed lawn clippings is to give them immediately and only in small amounts that will be quickly eaten, with none left in a pile. The only other option is to leave clippings on the lawn to dry completely before raking them up. Still, this is only safe if the grass dries immediately and there are no clumps that stay damp. Horses shouldn't be allowed to graze a pasture that has been cut with a rotary mower, because that type of mower prevents proper drying by wadding up the grass instead of laying it down flat.

GRAIN

Grain is often fed to horses if they need more energy and nutrients than pasture or hay can provide. This is generally the case with lactating mares and hardworking or young, growing horses.

★ ★ ★

Storage Basics

Protect grain from moisture, as well as from rodents and other animals that will eat some of it and contaminate the rest with feces and urine, which can spread several diseases. In a barn, store grain where horses don't have access to it, preferably in a feed room with a horseproof door and latch. Ideally, a grain room or bin should have a concrete floor for easy cleaning and to thwart access by rodents. Installing ratproof wire behind the walls and above the ceiling will also keep rodents away. Store the grain in bins or mouseproof containers rather than sacks: Large plastic garbage cans with secure lids work nicely, and they don't condense moisture as much as metal in cold weather. Always refasten the lids after scooping out grain.

USING A CHEST FREEZER

Storing grain in an old chest freezer that no longer works will protect it from rodents and moisture. Wash the interior and dry it thoroughly before using it, to make sure it's clean and free of mold spores that could contaminate it. A medium-size freezer will hold about ten 100-pound sacks of grain. Caution: Remove the latch so a child can never accidentally be trapped in the freezer!

Feeding Grain by Weight

Grain should always be fed by weight rather than volume, because some types are heavier than others. You may create digestive problems, colic, or founder if you feed by the scoop or by a certain amount in a coffee can and then change to a heavier grain without adjusting the volume.

Heavier grains not only take up less volume but they also have more energy density. For example, corn has almost twice the energy of an equal volume of oats. You can weigh a scoop or can of grain, but the weights listed here are good approximations.

GRAIN	# OF POUNDS PER QUART
Oats	1
Barley	1½
Corn	1¾
Milo	1¾

If your horse gets adequate energy from 5 quarts of oats per day and you suddenly feed him 5 quarts of corn, you're really increasing his feed from 5 pounds to 8¾ pounds. At that point, you're overfeeding him and putting him at risk for weight gain, colic, and founder.

First determine the correct amount of feed for your horse (how much he actually needs according to his age, weight, and the work he does). Next weigh the amount of grain that your coffee can, scoop, or ice cream bucket actually holds. Once you know what it weighs, you can easily adjust your horse's feed.

Because whole grains are denser, they weigh more per volume than cracked grains, which in turn weigh more than ground grains. For example, cracked corn is heavier than ground corn but lighter than whole corn. Wheat bran weighs less. The weight of commercial, processed feeds can vary a lot, depending on the type, quality, and processing method. A can of pelleted grain weighs more than a similar can of sweet feed containing molasses.

Jug Scoop

You can make a handy scoop from an empty gallon milk jug, liquid bleach jug, or any other sturdy plastic container simply by washing it out and cutting off the bottom with a sharp knife. Leave the lid tightened so the scoop won't leak.

Coffee Can Scoop

To make a durable scoop from a 2-pound coffee can, buy an inexpensive metal drawer handle and two short bolts with nuts. Position the handle near the rim and drill holes in the can to accommodate the bolts. When using the can as a grain scoop, weigh the can with feed in it (the type of grain you always use) at different levels of fullness, deducting the weight of the can to get accurate measures for the grain and marking them on the can.

Rubber-Tub Feeder

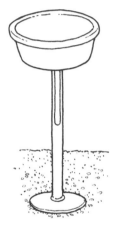

A rubber tub can hold water, grain, or salt, but horses will often paw at or tip them over. One way to immobilize a tub and get it up away from horses' feet is to mount it on a stand made with a vertical metal pipe. Even horses that like to chew on tubs will rarely dismantle this feeder because it takes a direct upward pull to lift the tub out of the base. And it's safe because a horse won't run into the metal pipe; he'd have to bump the flexible rubber tub first.

To build this style feeder, you'll need the following materials.

- ★ 2-inch metal pipe, 3 feet long
- ★ 1¾-inch metal pipe, 1½ feet long
- ★ Old metal disk of any size
- ★ 10-inch metal plate, round or square
- ★ 4 bolts
- ★ Rubber feed tub

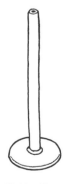

For the base weld one end of the larger pipe to the center of the metal disk. (If you don't weld, have a professional welder to do it.) Bury the disk end of the pipe where you want the feeder so that about 2 feet of the pipe is sticking up above the ground.

Weld a disc to the base.

For the top weld the smaller pipe to the center of the flat metal plate. Drill matching holes in the metal plate and the feed tub, then bolt them together. Once you slide the small pipe into the larger base pipe, the feeder is ready to use. To clean the tub, just lift it out of its base and dump it.

Bolt a feed
tub to the top.

Feeding Grain to a Group

In a pasture, it's best to have no more than 10 horses feeding together (the fewer, the better). Whether you put the grain tubs or buckets along a fence line for ease of filling or out in the open, make sure they're at least 30 feet apart so the bossiest horse won't chase the others away. Place an extra tub at a distance to give the most timid horse a place to eat. The more dominant horses generally eat from the closest tubs.

To prevent horses from spilling grain from tubs that are set on the ground, use ones designed to stay put. Some are equipped with a big slippery upper lip to make it difficult for horses to grasp with their teeth, and with a bottom outer lip that will catch the ground and keep the tub upright.

Feeding a Foal

A foal learns to eat grain by following his mother's example and sampling what she eats. But he eats more slowly than the mare and may manage only a few bites while she consumes the rest. To make sure the foal gets a certain amount of grain, you must tie the mare where she can't reach his tub or feed them in separate stalls.

Another option is to make a creep feeder. In a pen or the pasture, set up a pole or board at the entrance to the foal's feeding area so he can walk under it but the mare cannot. If you feed them in a stall, using feed boxes, place a series of small bars over the foal's box so his small muzzle will easily fit between them but the mare's won't. You can also use a small bread or loaf pan (one too small for the mare) with a handle or flange at each end. Snap the handles (or drill a hole in each flange so you can attach a snap) to the stall wall at the foal's height.

BUCKET HOLDERS

A simple way to keep a horse from overturning a feed or water bucket is to set it in a tire. It'll remain upright, even when horses root or bump into it. The tire is resilient enough to have some give, and it won't cause injury if a horse gets chased over the top of it by a pasture mate.

In a stall an old feed or water bucket that leaks makes a good elevated holder for another bucket in a stall; position it at the appropriate height in a corner and drive a screw in each side where it touches the wall. The holder allows you to remove the inner bucket to clean it out or fetch more grain or water.

FEEDING MANAGEMENT

A horse's diet is something that needs to be monitored and adjusted whenever there is a significant change in his activity level or physical condition. In addition, the availability of different feeds can depend on the seasons.

Hay versus Grain

Unless they're growing, lactating, or working hard, most horses receive adequate nutrition from good pasture, without additional hay or grain. If you have no pasture, good hay is usually adequate. If your horse gets both hay and grain, at least half the weight of the total ration should be good hay; horses need the roughage for proper digestion. Adult horses generally do fine on grass hay, but young growing horses and lactating mares need the extra nutrients from legume hay, such as alfalfa.

Feed a horse more hay, especially grass hay, at his evening feeding when he has more time to eat and digest it. The extra hay at that time will keep him busy, so he'll be less likely to chew on the stall or fences.

Feeding Schedules

The horse is a grazer whose digestive tract works best when handling small amounts of feed continually. Depending on the quality of the grass, a horse at pasture may graze for 12 to 17 hours every day. A horse confined to a stall or pen is usually fed only twice a day, and he falls into the habit of eating his large meals all at once with nothing to do in between. His digestive tract is alternately overloaded and empty.

Most horses adapt fairly well to this unnatural situation, but some tend to have bouts of indigestion or colic. Those individuals are better off feeding at more frequent intervals. They'll utilize their feed better and be healthier and happier. Even if you work or go to school and are not home part of the day, a three-times-a-day feeding program can work: early morning, after work, and at night before bed. For the average horse, this doesn't mean increasing the total amount fed, but splitting the daily ration into three portions instead of two.

Feeding a Pregnant Mare

As a mare enters her last trimester of pregnancy, the growing fetus takes up more space in her abdomen, leaving less room for the digestive tract. Consequently, the mare gets full faster and can't eat large volumes of high-fiber feed, such as hay. If she's on pasture, she can graze often, but if she's fed with other horses, she won't get her share. She may stop eating before they do. To resolve this problem, put her in a pen by herself and feed her small amounts more frequently.

EATING & WORK DON'T MIX

A horse should finish eating at least an hour before he does any hard work, although a small treat like a handful of grain as a reward for being caught is fine. After working, a horse can safely eat green grass or hay, but don't give him grain when he's hot and tired. To avoid digestive problems, let him cool out completely and wait at least an hour or two before feeding him.

Tempting the Finicky Eater

Keeping weight on a working horse or a mare that's nursing a foal can be difficult, especially when the horse is a fussy eater. With a horse that's losing weight, the first option is to increase his feed or to boost the energy density of his feed by adding grain or legume hay. A hardworking horse, however, may not eat enough hay and may need something more tempting and palatable, like fresh green grass. In fact, a tired, dehydrated horse will often eat green grass when he won't touch anything else.

Take care when increasing alfalfa or grain in a horse's ration, because too much of those rich feeds can cause colic or metabolic problems. Sometimes the best way to increase consumption is to feed many small meals rather than two or three large ones each day. If two or more horses are living together, the fussy eater may need to be separated so he can consume his share. Often a finicky horse will eat awhile and stop. If his companions are gluttons, they'll eat his share before he can come back for another snack.

Sometimes a few drops of apple cider vinegar added to his grain or water will encourage a horse to eat or drink. Most horses seem to like the tangy taste. Vinegar can also be used to accustom a horse to drinking unfamiliar water. If he becomes used to the taste of the water at home, he'll more readily drink strange water if you spike it with vinegar.

Feeding a Sick Horse

If a sick or recovering horse won't eat, you can force-feed him a few meals with a well-washed, large dewormer syringe. Have your veterinarian recommend the types of food most appropriate for your horse in his condition; mix those with a little water and molasses in a blender to create a thick soup or thin mush that can be given with the syringe. Some of the foods you may be able to give him are apples, carrots, raisins, yogurt, and cottage cheese. Pelleted feeds can be soaked in a little water until they fall apart, then stirred into an "instant dinner." A few ounces of dry red wine given by syringe twice a day will often stimulate his appetite.

Food given this way provides energy and nutrients until the horse starts eating again. Just insert the syringe into the corner of his mouth and give him the mixture as you would a dewormer. Then hold his head up and wait for him to swallow it before giving the next syringeful.

CHORE LIGHT FOR DARK NIGHTS

If you feed and water horses after dark, a standard flashlight is hard to manage when you need both hands free for using a hose, throwing hay, or carrying feed. A more practical choice is a hands-free flashlight, such as the type that coils behind your neck and holds the light in place beneath your head, or better yet, a head lamp that fits into a browband above your eyes.

FEEDING MANAGEMENT

Changing a Diet Gradually

When you make a change (for example, switching from hay to pasture or from grass hay to alfalfa hay, or adding grain to the diet), do it gradually. When changing from one hay to another, mix the two kinds for several feedings, adding more of the new hay each time, to give the horse's digestive system time to adjust. Quick changes can disrupt the pH (acid–base environment) of the large intestine, affecting the microbes that help in digestion. Suddenly feeding a rich legume hay may make the horse sick.

To change from hay to pasture, put your horse out only 20 minutes the first day, 40 minutes the second, 90 minutes the third, and so on until he's out full-time. This helps avoid digestive problems or grass founder. If your horse has foundered on grass before, either limit his grazing to an hour a day or keep him in a paddock and feed him hay.

Preventing Food Poisoning

Here are some feeding pointers for safeguarding your horse's health.

★ Always check feed for unusual odors and abnormal appearance or texture. Discard any suspicious feed or moldy flakes of hay (away from places horses have access to them).

★ Separate the feeding and watering areas so that the horse won't dribble feed into his water when he decides to drink; grain or hay can ferment in water, causing harmful microbes to grow.

★ Regularly clean feed tubs and waterers, including the corners of feed boxes or mangers where old feed may build up. Never leave uneaten feed in a bucket or tub for more than 12 hours. Moisture from the horse's mouth when eating may be enough to start the growth of molds and other fungi.

TREATS & SUPPLEMENTS

Just because food or fruit is nutritious for humans doesn't mean it's safe for horses: It's best to stick to apples and carrots. The safest treat of all (and one that any horse that's not on pasture will appreciate) is a bouquet of handpicked, long-stem green grass. In addition, certain feed supplements can provide valuable nutritional boosts.

★　　★　　★

Feeding Treats

Here are some things to keep in mind about feeding horses human foods.

★ Many horse owners like to give treats, such as carrots, apples, and sugar lumps. It's important to hand-feed treats carefully or a horse may become nippy and demanding. Most experienced horses can handle a carrot or an apple, but a young horse or one unaccustomed to this type of snack may try to swallow portions that are too large. To be safe, cut a large treat into smaller pieces.

★ Some people give horses cookies, soft drinks, and other human treats. Most are probably harmless, but others are not. Chocolate and some soft drinks contain caffeine, which is not good for horses. Chocolate also contains compounds that could show up on a drug test at a horse show and lead to disqualification.

★ Citrus rinds contain components that can be harmful when eaten in large quantities.

★ Pits from apricots, peaches, and cherries contain cyanide. A swallowed pit that passes through the digestive tract poses little danger, but a chewed pit releases toxic substances that can make a horse ill.

★ Potato skins contain solanine, a complex alkaloid found in nightshade plants. Horses are more sensitive to it than humans are. Other plants containing alkaloids that can give a horse indigestion are tomato, eggplant, and avocado.

★ Watermelon rind can cause choking if a horse tries to swallow large pieces without adequately chewing.

Feeding Salt

Horses always need salt, and it's best to provide it in a way that allows the horse to regulate his intake according to his needs. To keep the salt from melting, put salt blocks or boxes on high ground where they'll never be flooded. In a rainy climate, use a salt box with some type of roof that still allows the horse easy access. Make sure fresh clean water is always available; overeating salt won't hurt the horse as long as he has plenty of water to flush the excess in his urine.

Giving Liquid Supplements

An easy way to measure vegetable oil, wheat germ oil, or any other type of oil or liquid supplement is with a large syringe or a long, thin dewormer syringe. Draw the amount you want to dispense into the syringe, then note the level with a permanent marker. In the future, you can just stick the syringe into the supplement container and suck up the amount needed. Clean the syringe between uses and keep it next to the supplement.

STORING SUPPLEMENTS

Fat supplements or any grain or supplement containing added vitamins, such as A or E, will lose quality over time. Oxidation can turn fats rancid and degrade nutrients. Buy supplements in quantities that will be used quickly, and store them in a cool, dark place protected from air and sunlight.

WATER

Horses should always have access to clean water. They will drink more in hot weather or when working and sweating, or lactating.

★　　★　　★

Making Your Own Water Tank

For a water tank large enough to serve several horses in a pen or pasture, try cutting a 55-gallon barrel in half crosswise or lengthwise with a handsaw. You can probably obtain a used barrel from a bakery or a restaurant. Be sure the barrel was used to store or transport food products, not toxic chemicals; wash it thoroughly. If you cut the barrel lengthwise rather than crosswise, it will be less apt to move around or tip over if you set it in a custom-shaped indentation in the ground. You can lift it out periodically to dump and scrub it.

Water Tub and Tank Cleaning Tips

Here's how to keep your tank or waterer fresh.

★ Remove floating debris from a tank or waterer with a kitchen strainer.

★ If algae, moss, or grime accumulates on the bottom or sides of a tank or tub, scrub it off and rinse the container well before refilling it. A long-handled scrub brush, such as a clean toilet brush, works well.

★ An instant disposable scrubber can be created by wadding up used baling twine into a ball. The twine is abrasive enough to scrub even the most stubborn grime.

★ If nothing else is handy, use a handful of hay folded into a pad to scrub slime and algae. Be sure to discard the scummy hay afterward so the horses won't eat it.

Spillproof Water Tub

Some horses play with their water tub or bucket, pulling it up with their teeth and spilling it. If it's on the ground, they may paw it with their feet and foul the water with dirt and manure or flip the tub over.

To thwart the tub-tipping foot dunker, you need to move the tub off the ground; the horse won't be able to walk into it, spill it, or steal it. In a stall you can hook the tub to the wall. Outdoors, snap or tie the tub to the fence; create an attachment place by drilling holes in the rim of the tub and inserting eyebolts secured with washers and nuts.

Another solution is to set the tub snugly into an old tire, then, if needed, you can place place the tub-in-tire on top of another slightly larger tire to achieve the height necessary to keep him from pawing it. When the tub is full of water, it will be fairly heavy and difficult for the horse to pick up with his teeth. But when you need to rinse or clean the tub, all you have to do is pull it straight up and out of the tire. The tires are heavy enough to keep most horses from moving them around, yet resilient enough that a horse can't be injured if he runs into them. If a horse persists in trying to pull the tub out of the tires or move the tires around, put eyebolts in the rim of the tub and snap it to smooth wire you wrapped around the top tire, then fasten the tire to the fence behind it.

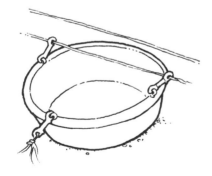

Tub secured by eyebolts

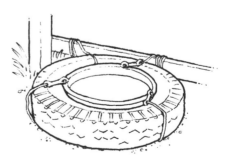

Tub-in-tire

No-Spill Water Carrier

Plastic water containers with spouts and lids are handy, but you may not always have one available. When you're hauling water in buckets, one way to make sure you don't accidentally spill it is to put a plastic trash bag into the bucket, fill the plastic bag with water, and then tie a knot in the top of the bag. The water can't splash out, even if the bucket tips. You can haul water this way in a wheelbarrow, the trunk of a car, or the back of a pickup. Once you arrive at your destination, just untie the bag, fold the top of the plastic bag down over the sides of the bucket, and pour out the water or let the horse drink from the bag while it's in the bucket.

Line the bucket with a trash bag.

Tie the bag securely once it's filled with water.

Watering in Winter

Horses may not drink much in cold weather, but they'll always drink a little when eating. It's not imperative that the horse have unfrozen water all night long (he won't drink during the coldest hours), but he does need drinkable water during his evening meal. If you make sure he has an adequate supply through the day and evening, he won't become dehydrated. Ideally, the water should be no colder than 45°F when you give it to him; he may not drink enough if it's too cold.

Preventing Ice in a Water Bucket

During cold weather, a horse's water tub or bucket may freeze over, and if the ice gets thick it's hard for a horse to break unless he puts his foot in it. One way to keep ice from forming so solidly is to place a partial lid over the tub or bucket, leaving just enough room for the horse to fit his nose in and drink. The lid can be made from plywood or any other durable material secured to the top of the container with snaps. Even if the water still freezes, the ice won't be nearly as thick as it would if it were uncovered.

Solar-Heated Bucket

Another idea for keeping water from freezing in a bucket is to use two old car tires with an inner diameter that will accommodate the bucket. Stack the tires, then wire or bolt them together. Fill the insides with sand or fine gravel. The black rubber will absorb heat from the sun during daylight hours, warming the sand or gravel. It will take a while for the heat to dissipate after sundown, even on a cold night, allowing the horse to have drinkable water with his evening meal.

Insulating a Water Tank

A 32-gallon plastic trash can makes a good water tank (it's safer for horses than a metal one) when set in the center of two stacked tractor tires. If the inside diameter of the tires is too small, use a chain saw or jigsaw to widen it; you want the trash can to fit with 2 inches to spare. Fill that space with fiberglass insulation or Styrofoam, but not all the way to the top where the horses can get at it. Unless the temperature drops extremely low, the tires and insulation will keep the water from freezing.

Another method is to stack three large truck tires and fill the insides with straw for insulation. Then set your water drum inside the tires with another inch or so of straw on the ground under it. The black tires will absorb heat during the day, and the insulation will help keep the water from freezing at night.

CHAPTER TWO

The Barn

Whether your barn is large or small, simple or elaborate,
there are always ways to make it better and safer
as well as to simplify many barn chores.

BARN CONSTRUCTION

If you already have a barn, it may merely need some maintenance or repairs. If you are building a new barn, design it to best suit your purposes and to be safe for your horses.

★   ★   ★

Making It Last

When building a new barn or remodeling an old one, don't scrimp. Cutting corners and using cheaper materials may end up hurting a horse or a person if something wears out or breaks. Buy good doors with good hardware, and build stalls with materials that are safe and will endure.

Being Weather-Wise

Take the elements into account when you plan a barn.

★ For a large barn, allocate the south side for stalls and grooming areas and the north side for an office, the tack room, and hay storage. This creates a buffer between the horses and cold northerly winter winds, and puts the horses on the shady side during hot summer afternoons.

★ In a snowy climate, place doorways on the sides that are not directly under the slope of the roof. If snow slides off the roof, the pile won't obstruct the door.

★ Install overhangs to shelter doorways and keep them free from puddles and mud.

★

STRENGTHENING METAL STALL WALLS

Lining a metal wall with wood will keep a horse from kicking through it and cutting his leg. A solid lining is always safer than a few kick rails or boards; a horse could catch his hoof between them.

Filling & Covering Building Blocks

If you're planning to build with cinder or concrete blocks, be aware that they're not as durable as they seem. Because portions of the blocks are hollow, horses can kick through them and be injured by the sharp edges. When using blocks for stall walls, fill the hollow parts with sand; also line the stall interior with wood from the floor to a height of at least 5 feet. A smooth wood covering will keep the horses from scraping themselves on the rough blocks.

BEEFING UP WALL INSULATION

Some barns don't have enough insulation to keep the interior warm in winter and cool in summer. Stapling empty plastic-weave grain sacks to the barn walls is an easy way to add some insulation.

Installing Wood Stall Walls

Running the boards vertically is usually safer than arranging them horizontally, because they'll be shorter and stronger. Long horizontal boards need vertical reinforcements, spaced no more than 4 to 6 feet apart, to keep them from bending or separating if a horse kicks or pushes on them. Without the supports, he may accidentally lodge his foot between the boards and seriously injure himself.

Removable Stall Divider

A removable wall lets you turn two stalls into one large one. This can come in handy for foaling, for a large horse, for a horse that tends to roll and get cast (can't get up) in a small stall, or for a sick or injured horse that has to stay in the barn but needs room to move around.

An easy-to-build divider consists of horizontal planks that slip in and out of grooves in the barn walls. To form the grooves, use channel iron or wood braces made from 2 X 6s lag-bolted in place. A horse won't be able to move the planks if you screw vertical strips to the divider, creating a solid wall.

TIPS *for*
PREVENTING HORSES FROM CHEWING WOOD

Try to stop chewing before it starts. Once a horse gnaws a particular spot, he finds it easier and more enticing to continue.

★ Commercial products to halt wood chewing have a foul taste and will dissuade most horses, but are not harmful if ingested. Homemade recipes can also be effective; try the ones listed on page 80.

★ Consider eliminating edges and corners when planning a stall with a wood interior.

★ An inexpensive way to protect stall partitions, doors, and posts is to cover the edges with aluminum corner moldings used in house construction. They're easy to install and to keep clean. Cut them to the length needed, or put several end to end. Drill screw holes in the aluminum and sand any rough edges before installing the moldings.

Ventilating Stall Fronts and Doors

Most stalls have solid walls and doors that trap ammonia inside the stall. When building or remodeling, consider using a 12-foot tubular metal gate instead of a prefabricated stall front. Not only will it cost less and provide better ventilation, but it can be opened wider to allow equipment into the stall for cleaning. It also increases visibility for the horse looking out of the stall, and for you looking in. If you decide to put in a different stall front later, you can use the gate elsewhere.

Slatted or mesh stall fronts and doors made of metal gratings also provide wonderful visibility and ventilation. A stall front of steel mesh with spaces about 1¼ inches wide will keep even a small foal from poking a foot through. If the stall front is made of pipe or metal bars, space them about two inches apart so a horse can never catch his nose, teeth, or lower jaw when trying to nibble or chew.

DOOR HARDWARE

Make sure all latches, handles, and locks are functional from both sides of the door and that they don't protrude so much that a horse passing by or rubbing can snag himself on them. You also don't want to make it easy for a clever horse to open them; see page 78 for foolproof latches.

Sliding or Swinging?

Stall doors should be at least 4½ feet wide and 8 feet high (preferably as high as the ceiling). The traditional Dutch door is still popular, as it enables a horse to stick his head out when the top part is open. Sliding doors are the safest, though, particularly with wood on the bottom and grillwork on the top. Some doors have grillwork that folds down, letting the horse poke his head out into the aisle when you're in the barn.

Avoid swinging doors, if possible. They can be a hazard if they swing into the stall, because a horse eager to come out may bang into the door before you open it completely. Doors that swing outward pose a hazard to passing horses and humans.

ADVICE *on*
WIRING & LIGHTS

To work efficiently in the barn you'll need good lighting and plenty of outlets.

★ When building a new barn or rewiring an existing one, have an electrician do the work to make sure everything is properly grounded.

★ Locate light switches next to the main doors so you can turn them on as you go into the barn.

★ For ample light in each stall, use a 150- or 200-watt bulb or a fluorescent light with a minimum of 40 watts. To increase the amount of light, position the fixture near the front of the stall (high out of the reach of the horse), and use it in conjunction with a reflector.

★ Protect stall light bulbs with a covers to prevent any fragments of a shattered or exploding bulb from dropping into the bedding.

★ Clean cobwebs from all fixtures to reduce fire hazard.

★ Locate a grounded outlet with a cover outside every stall so that you can unplug something without having to go into the stall. If the stall front or door allows the horse to poke his head out, place the outlet very low or off to the side where he can't reach it.

Windows and Screens

A window in each stall will brighten the barn and (if it can be opened) improve ventilation. In a warm climate grillwork or bars over the window openings may be enough. In a cold climate you can use shutters or sliding acrylic plastic windows outside the bars or grillwork instead of glass. If the stall has a glass-paned window, install a durable barrier in front of it that still allows you to open the window but prevents a horse from touching the glass and breaking it.

Floor Options

Stall floors should drain well and provide good traction. Neither wood nor smooth concrete is a good choice, because moisture can't drain through and the surface is slippery. Mats laid on the wood or concrete will improve traction, but they won't solve the drainage problem.

The best floors are porous and have some give. Dirt or gravel topped with bedding provides excellent drainage as well as cushioning when a horse lies down. Unforgiving surfaces may cause pressure sores. A resilient floor is essential for foals, which tend to develop sores on the outside of their hocks when they scramble up from a hard floor.

Traction is just as important in the aisle as in the stalls; the surface should never be slippery. A dirt floor, a highly textured concrete or asphalt floor, or textured pads or mats are the best choices.

Buying Stall Mats

Easy-to-clean stall mats are a good solution for stall floors that are slippery, dusty, or hard to keep free of holes. They can also save you money; because of the cushioning they provide, they need less bedding over them than a bare floor does. Here are some factors to keep in mind.

★ Stall mats are available in many materials: rubber, plastic, or a combination of rubber and synthetic materials. A textured surface offers the most traction.

★ Some mats are designed to let moisture through, which works well on a dirt or gravel floor but not a nonporous floor.

★ Soft rubber provides the most cushioning, but it may wear out quickly. The denser the rubber, the more durable the mat.

★ Tile-grid flooring, which comes in small, interlocking squares, is portable. You can take it along when traveling, to put down in any stall your horse uses.

THE VALUE OF ULTRAVIOLET LIGHT
Windows and skylights make barns brighter and healthier. Ultraviolet light in sunshine kills many airborne viruses and bacteria, and even some parasite eggs and larvae. Light is also essential for normal body functions, such as reproduction in mares and shedding winter hair in spring. For skylights, choose plastic or translucent glass that will allow most of the ultraviolet light to penetrate. If your barn has a metal roof, you can incorporate translucent fiberglass sections.

Making Your Own Stall Mats
You can fashion stall mats from a discarded conveyor belt, often available at lumber mills, stone quarries, and gravel crushers. With a carpet knife cut the rubber belt to fit, and lay the strips side by side. The big strips are also handy in cross-tie areas, aisles, washrooms, and any place where you want a secure walking surface that can be easily washed.

Designing a Wash Stall
A wash stall needs good drainage as well as good footing, because the floor will be wet and some horses become fractious or upset when bathed. An asphalt surface is not as slippery as concrete, but a concrete floor will work if it's textured (finished with a bristle broom) instead of smooth when you pour it. Some types of rubber mats placed over the flooring will work well, too. Design the stall so the water source is handy, but the taps are recessed or protected to keep the horses from running into them. Place the drain at the back of the stall so the horses won't have to walk over it.

VENTILATION

Good air flow helps minimize dust, ammonia, humidity, mold spores, and other unhealthy conditions in a barn. Here's how to evaluate your air flow as well as ways to improve it.

★　★　★

Checking the Air Exchange

You can assess the air flow in your barn by closing all the windows and doors and walking through the entire place. If it's built for good ventilation, there will still be some air movement. One way to better assess this is to use a smoke puffer, a small device that fills the barn with harmless smoke. You can see where it goes and monitor how fast it dissipates. You'll find out exactly which areas of the barn have stagnant air. For good air quality in a barn, you need a minimum air exchange rate of about six to eight complete air changes per hour. (One air change is the time it takes for the smoke to completely disappear from the barn.)

Creating Ceiling and Floor Air Flow

In some barns, just opening the windows, doors, and tops of stall doors gives you good ventilation. Interior walls that don't reach clear to the ceiling allow warm, stale air to escape. Other ways to improve air flow include putting in vented openings at the bottom of a stall wall to draw in fresh air; installing a vented opening at the bottom of the exterior wall with a matching vent at the bottom of the stall door to increase floor-level ventilation; and replacing solid stall doors with grills.

Installing Fans

In some cases you may need a fan to force air through the barn, but just make sure you don't move so much air that you create a draft.

★ Large fans in the gable at each end of a pitched roof can help with ventilation.

★ A paddle fan above each stall (near the aisle door) will circulate air through the barn, as well as keep horses cool.

★ An exhaust fan or two, near the ceiling, will keep air from becoming stagnant, even when the barn doors are shut in cold weather.

★ To move the most air, use fans designed to pull air up, not blow it out.

★ Set fans on timers, for example, to run for two minutes every hour.

Preventing Humidity

Dampness in barns is caused by the evaporation of moisture from urine and manure, and all the tiny droplets of water in the air exhaled by horses. The average 1,000-pound horse exhales about 2 gallons of moisture daily. Another source of dampness is condensation on the underside of a metal or uninsulated roof dripping into the barn in cold weather. Besides damaging the building and leading to feed spoilage, damp air (coupled with irritation of air passages from ammonia, dust, and mold from hay and bedding) will contribute to respiratory problems.

To prevent high humidity in a barn, you need to vent damp air out and allow fresh air to replace it. Don't close all the openings in winter to keep the barn warm. This restricts air movement too much and traps moisture and ammonia fumes inside. Instead, reverse the direction of the ceiling exhaust fans (if you have them) to pull in cold air to mix with the warm air in the rafters, causing it to sink and warm the barn. The mixed air can then be expelled through adjustable ground-level louvers, or even through burlap-covered windows. Cold fresh air is always healthier for horses than stuffy, humid air trapped inside a barn that's been tightly sealed to preserve heat.

SUGGESTIONS *for*
VENTING THE ROOF

Here are ways to move out the warm air that rises and collects in the rafters.

★ When building, keep in mind that roofs offering the best ventilation are steep rather than flat, with about 6 inches of rise for every 12 inches of width.

★ An easy, cheap way to improve ventilation is to install metal ridge vents (about 10 feet long, with an opening about 6 inches wide) along the roof peak. The size of the barn determines how many ridge vents are needed; a 100-foot-long barn, for example, needs four vents. You can augment the ridge vents with a spinning ventilator.

★ A ridge cap allows warm air to move out of the attic space, though it can't move as much air as the taller ridge vent. It's often combined with other devices, such as cupolas and louvers. A cupola, usually four-sided with louvered openings or windows on each side, acts as a chimney to pull hot air out of the barn. You can install louvers at both ends of the barn beneath the roof peak; they work best when the roof is steeply pitched.

KEEPING STALLS CLEAN & SAFE

To ensure that your horse's stall is a healthy, safe, and enjoyable environment for him, it's important to choose the right bedding, feeding, latches, and toys.

★   ★   ★

Bedding Options

Many kinds of materials make good bedding; the main factors to consider are cost, availability, amount of dust, and palatability. You do not want something the horse will eat. Choose a material that is compatible with your stall flooring and that you can easily dispose of.

★ Straw is absorbent and provides good drainage; the soiled bedding can be composted or spread in a field. Wheat straw is softer than barley straw and less palatable than oat straw.

★ Wood chips or shavings aren't as dusty as more finely ground wood, though some chips are too coarse and sharp. Shavings make a soft bed that horses won't eat, and some softwood types are even more absorbent than straw. However, don't use wood shavings to bed foaling mares, because wood products can harbor bacteria that may harm the newborn. Some hardwood shavings can be toxic to horses: Black walnut causes laminitis, oak is too acidic, and yellow poplar can cause itching. All types tend to dry out a horse's feet, and some are dusty. Compost used shavings before using on fields or pastures.

MAKING STRAW BEDDING UNPALATABLE

If a horse insists on eating his bedding, you can discourage him by very lightly spraying Pine-Sol cleaner over it. The smell will deter even the most avid straw eater.

★ Volcanic aggregate (lightweight and porous) can be layered under other bedding materials to provide better drainage.

★ Shredded paper is very absorbent and fairly dust free; it's comfortable for the horse and easy to clean out of stalls. Ink in newsprint or phone books is a nontoxic vegetable dye, but some other types of paper may not be as safe (check the source before using it). The carbon in ink tends to absorb and reduce odors, but the ink itself may rub off on light-colored horses. Even when horses eat some of the paper, it doesn't seem to harm them.

★ Rice hulls work best in stalls that drain well, or in conjunction with another type of bedding, such as wood shavings. Because the lighter rice hulls rise to the top and the heavier, more absorbent shavings settle to the bottom, the mixture acts in the same way as a multilayered disposable diaper: Moisture seeps through the hulls into the more absorbent shavings, leaving the top surface dry and comfortable. Rice hulls also have the advantage of being less flammable than other types of bedding.

BEDDING ABSORBENCY

The following list ranks bedding materials from the most absorbent to the least absorbent.

1. Peat moss
2. Shredded newspaper
3. Pine sawdust and chips
4. Long oat straw
5. Wheat straw, pine shavings, and barley straw
6. Rice hulls and hardwood chips or shavings

KEEPING STALLS CLEAN & SAFE

Keeping Stalls Dry

Several commercial products are available to help absorb moisture and odor from urine and manure. Lime is traditional and works well, but it's caustic and can irritate a horse's eyes and nostrils and dry his hooves. A relatively inexpensive alternative is powdered marking chalk (available in 50-pound bags from lumberyards). It's less caustic than lime and works just about as well. Another option is kitty litter, which is easy to sprinkle on stall floors and is safe for horses. A handy way to spread lime (if you use it) or powdered chalk is to sprinkle it through an old flour sifter or kitchen colander. That way, you can spread it evenly with no piles or bare spots.

Minimizing Ammonia

Heavier than air and hanging low near the floor, ammonia fumes are a major cause of lung irritation in foals, which spend a lot of time lying down. Adult horses can be affected, too, because they lower their heads when they eat hay. You can get a good sense of the air quality down there (the highest concentration of fumes is within 3 feet of the floor) by sitting or lying on the stall floor. If you smell ammonia gas, it's already twice the concentration at which it becomes harmful. The levels are often the greatest at the center of the stall, just above the bedding surface.

Keeping stalls clean and improving ventilation can help reduce fumes. Straw bedding soiled with manure and urine tends to produce more ammonia gas than does similarly soiled sawdust or shavings. If you use straw, clean the stall at least once a day, and preferably twice, to reduce ammonia buildup. You can use commercial products, such as Stall Fresh and Sweet PDZ, instead of lime. They are both byproducts of the mining industry and will reduce odor and dry out wet spots in a stall, trapping ammonia and eliminating the gas. Sodium bisulfate also helps control ammonia, as does a liquid product created from the yucca plant. An application of any of these products (sprayed or sprinkled on the cleaned floor before rebedding it) can be effective for up to three days.

When you don't have time to thoroughly clean a stall, spreading lime, chalk, kitty litter, or one of these products over the wet areas and covering it with fresh bedding can help keep ammonia levels down until the stall can be properly cleaned.

Rescuing a Cast Horse

Sometimes when a horse is in a stall or a small pen, he'll roll or lie down too close to the wall or fence and won't be able to position his legs underneath his body to push himself up. For example, a horse with colic may roll because of pain, oblivious to problems presented by the wall or fence until he's stuck against it. In horsemen's terms, he is cast. A foaling mare is another candidate for trouble. In some cases, a horse may end up on his back against the stall wall. In a pen, he may catch his feet in the fence rails, wire or netting, and you may need wire cutters or a crowbar to remove a fence rail or pole to free him.

Some horses in this predicament will lie quietly, waiting for help, but most will panic and struggle, beating their heads on the floor or ground or injuring their legs in the fence. If a horse struggles violently, he may twist an intestine. Thudding or crashing sounds from a stall, the beat of hooves against a wall, grunting, or horses whinnying (because their buddy is in distress) may signal the need for an emergency rescue. It's not difficult for two or more people to roll a horse away from a wall or fence, but you can do it by yourself, if necessary.

Walk calmly when approaching a cast horse, and talk to him quietly and soothingly. He's already upset, and you'll make the situation worse if you run, yell, or become frantic yourself. Approach his head quietly and in a low position, not only to avoid flailing hooves but also to alarm him less; a horse is instinctively afraid of something towering over his body.

If the horse is a short distance from the wall, he may just need to be pulled back a few inches to have enough room to get his feet back down so he can scramble up. Pulling his hindquarters by the tail may work, but often it's his front end that must be moved. A rope around his neck may give you enough leverage to pull him those few inches.

Loop a rope around a front pastern.

A horse that's clear up against the wall, however, must be rolled over by pulling on his front legs (which are up in the air), especially the leg nearest the wall. Start by looping a rope a couple of times around a front pastern, which will allow you to stay away from his flailing feet. Because he may be thrashing violently, do not try to grab his legs.

Brace a foot on the horse's neck.

Instead, brace one foot high on his neck, just behind his head, and push his head toward the wall with that foot as you pull his foreleg toward you. This will tip his weight toward the center of the stall, and he'll only need a little leverage to push himself back over.

Move out of the way.

As soon as he starts to come in your direction, move past his head and away from his thrashing legs. He will lunge to his feet as soon as he rolls away from the wall or fence, and you don't want to be in his way.

Once your horse is on his feet and has calmed down, check him for injuries, such as scrapes and cuts or dirt and bedding in his eyes. If he's in a stall, walk him outside and see if he's lame. Observe him for signs of colic, in case that was the reason he was rolling.

A horse that continually becomes cast may need a bigger stall or extra bedding around the edges of the stall to discourage him from lying near the wall. In his pen, place old tires along the fence where he usually gets cast to keep him from lying so close.

Keeping a Stalled Horse Entertained

Bored, confined horses may paw or chew on stall doors or fences, spill their water, or play with latches or anything else within reach. Some confined horses will develop stall vices, such as cribbing, weaving, or stall walking. If a horse can't tolerate confinement, give him something to play with so he can expend his energy on the toy rather than try to destroy the stall or develop bad habits or compulsive behavior. Try one of these ideas, keeping in mind that if your horse becomes bored with one toy, you can replace it with a different one.

★ A traffic cone, durable rubber ball, worn-out tire, rubber tub, or even a squeaky dog toy may divert him. He can chew it, paw it, pick it up in his teeth and throw it around, root it around with his nose, and otherwise entertain himself for hours.

★ Hang well-washed old plastic jugs, such as ones that contained milk or bleach, from the ceiling. Fill some with small pebbles or rice to create rattling toys. You can also suspend several jugs on doubled baling twine from the ceiling or wall, giving your horse something to rub on.

Using a Monitor

A baby monitor hung in a stall out of the horse's reach can be a real benefit when you're waiting for a pregnant mare to foal or tending a sick horse. If you keep the receiver with you in the house, you'll be able to hear unusual sounds, such as rolling, groaning, grunting, or pawing, and then head to the barn when the horse needs you.

FEEDERS & WATERERS

In every stall, you'll need an area where grain and hay can be kept clean. You'll also need a good and dependable source of water.

★　　★　　★

Safe Feeders

Corner hay feeders and mangers take up little room and allow horses to eat with their heads low, which is most natural. They also keep a horse from dragging hay around the stall where it will be tramped and wasted. If you're worried about a horse catching his foot on a wood feeder, use an inexpensive molded plastic model. Molded plastic is smooth but tough enough not to break or to hurt a horse that bumps into it.

Handy Grain Drawer Feeder

A sliding drawer at the front of each stall is a great time-saver, because you don't even have to enter the stall to give your horse grain. Old drawers from a remodeled kitchen work fine; use ones that will open just enough for a horse to eat from, but won't stick out too far into the stall. When feeding, simply walk down the aisle and pull open each drawer to pour in the proper amount of grain.

Water Bucket Holder

A plastic milk or delivery crate (often discarded by grocery stores) makes an inexpensive, sturdy, spillproof holder for a water bucket. Find one the proper size to hold a plastic 5-gallon bucket (you may be able to obtain a bucket or a similar-size pickle container used from a restaurant). Using a staple gun or hammer and fencing staples, attach the crate to the stall wall at the proper height for the horse. The hard plastic is resilient enough to prevent injury, and horses won't chew on it. If the plastic starts to crack, replace it with another crate.

Freezeproofing Water Pipes

In a cold climate, heat tape can help keep aboveground water pipes from freezing. Be sure to buy the kind intended for water pipes. Plug the heat tape into a grounded, three-prong outlet protected by a ground fault circuit interrupter (GFCI), so that a short won't start a fire or cause an electrical shock. Depending on the brand you buy, either run the tape along the underside of the pipe or wrap it around the pipe, then secure it with electrical tape or cable application tape. Be sure to locate the heat tape's thermostat at the coolest end of the pipe.

Following the manufacturer's directions, use insulation to cover the pipe and all the fittings and joints as well as the heat tape's thermostat (so it will register actual temperature around the pipe instead of the outer air temperature). Several types of insulation can be used, including polyethylene (a black tube 3 to 6 feet long, split lengthwise), fiberglass that comes in a bandagelike roll, and foam. Don't put it on any thicker than recommended or the heat tape may become too hot.

Insulating a Hydrant

If you have an upright water hydrant next to your barn that is not self draining or insulated, one way to insulate it is to put a 55-gallon barrel (with both ends cut out) over it and then pack the barrel with a well tamped mixture of straw and manure. Leave room at the top to fit a bucket under the hydrant or attach a hose. This should keep the ground at the base of the pipe as well as the pipe itself from freezing unless the weather turns very cold.

HEAT TAPE ON PLASTIC PIPES

When using heat tape on plastic water pipes, first wrap the pipe with aluminum foil to help distribute the heat more evenly. Never put heat tape on a plastic pipe or a water hose that doesn't contain water; the pipe or hose might become too hot and melt.

FEEDERS & WATERERS

HINTS *for*
STALL WATERERS

Maintaining a ready source of water is essential. In some instances that involves keeping pipes and filled buckets or tanks from freezing.

★ If you have automatic waterers, you may need to keep your barn warmer in cold weather. Even if the waterer itself has a heating unit, you must ensure that the pipes leading to it don't freeze. The underground waterline should be below the frost line, and the upright pipe to the waterer very well insulated or covered with heat tape (see page 47).

★ Position water pipes so that horses can't get to them. If a horse pulls a pipe off the wall in the middle of the night, the stall may be soaked by morning, and your well depleted. One advantage to having the waterer at the front of the stall with the pipe on the outside wall (in the aisle) is that it will be easier to check.

★ In a cold climate, the simplest solution may be to install a hydrant that won't freeze, the type that drains back underground when turned off, leaving no water in the upright pipe. Carrying pails of water or using a hose that you can drain afterward may be better than relying on automatic waterers that sometimes freeze or malfunction.

★

TACK ROOMS

Rather than buy a saddle rack and various gadgets for hanging bridles or holding grooming supplies, you can make those items yourself. It's a wonderful way to save money, recycle materials, and inject some personality into the tack room. A few of the projects described here call for welding; if you don't weld, you can always have your farrier or a professional welder do it.

★　　★　　★

Multisaddle Rack

A rack that lets you stack saddles is helpful when you have very little storage space in your tack room, barn, or house. This type of rack will fit anywhere there's at least 30 inches of unused wall space, even behind a door. Just be sure to position the rack where you can bolt it into a wood stud.

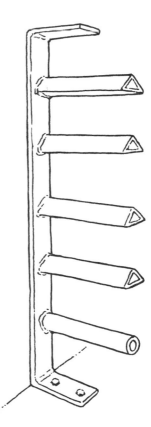

Start with a 5-inch-wide strip of steel (³⁄₈-inch thickness works well), cut to the exact height of the wall. Weld a 1-foot piece to the bottom at a right angle (to sit on the floor) and a similar piece across the top (to fit against the ceiling).

Next weld on the saddle rack "arms"; you can use large-diameter metal pipe (2½ or 3 inches), or 3½-inch angle iron positioned so the rounded ridge is at the top and the sides slope down. A typical ceiling height will allow for five or six saddle arms. To hold a large Western saddle, an arm should stick out at least 22 inches from the upright steel frame.

Secure the finished rack solidly to the wall, ceiling, and floor with lag bolts. Finally, pad each metal arm so a saddle won't rest directly on the pipe or angle iron. You can use whatever material you have on hand, cut to fit and secured underneath with duct tape. A quick and easy option is to use old wool socks. The wool padding makes a nice cushion for the saddle, and it also covers the end of the pipe or angle iron so there are no sharp or abrasive surfaces to scrape the saddle when you pull it off. Even socks with large holes in them will work fine if you layer several on each arm, making sure the holes don't overlap.

Milk Can Saddle Rack

You can make another type of "antique" saddle rack from a milk can with four lengths of metal pipe welded to the sides for legs. For stability, weld iron rods between the pipe legs. The can interior makes a handy storage spot, which you can close up with the lid, if you want.

Post-and-Pole Saddle Rack

If you have plenty of space in your tack room and want a rustic sawhorse-style saddle rack, you can make one from a large-diameter post (8 to 12 inches across) for the body and four poles (3 inches in diameter) for the legs. The number of saddles will deter-

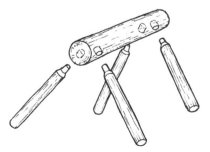

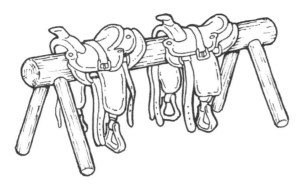

mine the length of the post; the one shown here holds two saddles, but it can be built to hold as many as four. Once you've cut the poles to length for the legs, whittle the tops to fit into holes drilled in opposite sides of the post at each end.

Nail Keg Saddle Rack

If you can find one, a wooden nail keg makes a nice rounded "antique" saddle stand that lets you store brushes and other grooming tools inside. One option is to secure the keg to the wall with a few long nails; attach a short length of 2 X 4 to the wall directly under the keg to help support it, and secure a longer 2 X 4 to the floor to prop up the front end. Another option is to attach the keg to an angle bracket fashioned from 2 X 4s nailed to the wall. A third alternative is to construct a freestanding movable saddle rack from a nail keg, using 2 X 4s and scrap boards to construct a supportive base to set it into.

Option 1

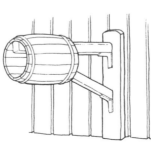

Option 2

Option 3

Horseshoe Hanger

Used horseshoes fastened to the wall with horse-shoe nails make nice-looking hangers for halters, bridles, and other tack. To form a hook, cut a horse-shoe in two at the toe, then weld half of that shoe at a right angle to the toe of a whole shoe. You can position the welded shoe with the heel up or down to create two styles of hangers. Either way, when the whole shoe is nailed to the wall, the upward-curving half shoe makes a perfect hook. Several of these hangers (painted or left natural) at various heights will provide maximum storage in a small area. The hangers are a great way to recycle old shoes, as well as display any unusual ones.

Jar and Coffee Hangers

You can create a hanger from a small jar, such as a baby food jar. Just nail the lid to the wall, then screw the jar back onto the secured lid. A rounded hanger like this is easier on a bridle crownpiece than a narrow hook is. Another simple bridle hanger can be fashioned by nailing coffee cans to a board, and then fastening the board to the tack room wall.

Wood Hanger

To make basic round wood hangers for bridles and halters, saw several 2- to 3-inch cross sections from a peeled post that's about 5 to 6 inches in diameter, then drill two screw holes through each slice. If you like, paint or varnish the wood slices before screwing them to the wall.

Grooming Tool Holder

A handy place to put your grooming tools is in an extra-large can set at a right angle to the wall and secured by screwing it through the bottom. You can also hang a bridle or halter over it. Another option is to hang a closet shoebag in your tack room or horse trailer and fill the pouches with various items.

IDEAS *for*
USING HANDY RECYCLABLES

Many kinds of empty household product containers can be useful in a barn, tack room, or trailer.

★ Coffee cans are handy for feeding grain. They're also ideal for storing grooming mitts or rags used for applying fly repellent (because the plastic lid keeps the cloth from drying out).

★ Plastic ice cream containers (3- to 5-quart size) with lids are useful for taking grain to a horse show.

★ Clean, empty spray bottles can be used for applying fly repellent and certain medications.

★ Plastic dish-soap bottles are great for storing many types of liquid. Often the squirt top will fit on the bottle of a product you plan to use, making it easier to administer.

★ Large trash bags with drawstrings make good saddle covers when you're riding in the rain.

PEST CONTROL

Rodents, insects, and birds can pose problems in a barn, but there are a number of effective ways to deal with them.

★ ★ ★

Keeping Out Rodents

Rats and mice can do a lot of damage in a barn or tack room. They eat feed and contaminate it, chew the twines on hay bales, and carry disease. They often enter through openings in vents, under doors, and around water pipes. Here are some strategies for keeping rodents out of the barn.

★ Check the walls and flooring to make sure there are no cracks or separations (such as in corners) allowing rodents to slip through. Seal any holes with sheet metal, steel wool (which can be pushed into any odd-shaped cracks), screening, or concrete. Also make sure windows and doors close tightly.

★ Clear debris, tall grass, and weeds from around the building. This type of cover attracts rodents, because they don't like to travel in open space where they are more vulnerable to predators, such as a good barn cat.

★ Store firewood, lumber, and other building supplies well away from the barn and at least a foot off the ground to minimize potential nest sites.

★ Remove debris that could be used as nesting areas, such as old newspapers or cardboard boxes.

★ Keep all feed and garbage in metal or thick plastic containers (that rodents can't chew through) with tight-fitting lids.

★ Sweep up any spilled grain or chaff and dispose of it well away from the building.

Poisoning Rodents

Apply poison that is labeled for use on rodents on or near their regular runways or entry holes, but only in areas where no young children, pets, or horses will have access to it. Always read the product labels and follow the directions, especially the safety precautions. When using poison, you must outsmart rodents so they'll eat enough to kill them. Limit their access to other food or water, so they'll be more likely to go for the bait. Check the bait every few days; replace what has been eaten and discard any that has become wet or dirty. Be sure to collect and burn any dead rodents you find to prevent your cat or dog from eating them and being harmed by the poison.

Here are some things to be mindful of when selecting a rodent poison.

★ Grain-based baits in meal or pellet form are effective placed along runways.

★ Rodent blocks, which have a wax base that rats and mice like to gnaw on, work well nailed to vertical surfaces along rafters or other off-the-ground rodent runs.

★ Some poisons are available in a water-soluble concentrate for making a liquid bait to attract rodents that have plenty of food available but no water source.

★ Tracking powders, applied with a squeeze bulb in thin layers along a runway or in a bait box, stick to a rodent's fur or feet; the rodent eats the poison as it grooms itself, or carries it back to its nest, where it may kill other rodents, as well.

Trapping Rodents

Rodent traps, such as the inexpensive wood snap trap and the flip box (which tosses mice into an escape-proof container when they pass over a triggering mechanism or enter a hole in the side of the box), pose less risk to pets and children than poison. However, they must be checked daily and the snap trap reset or rebaited when necessary. Although mice are easier to catch than rats because they're less suspicious and more readily attracted to traps and bait stations, you can also catch rats if you're persistent and tricky. Here are some pointers.

★ Place several traps where you see droppings and other evidence of rodent activity. Check them frequently.

Set the trap at a right angle to the wall.

★ Because a rodent uses its whiskers to follow the wall, it's more apt to feel its way around a trap that's parallel to the wall. Therefore, set the trap at a right angle to the wall.

★ Use wire to secure traps to overhead rafters, beams, or pipes when you're trying to catch roof rats or pack rats.

★ Place snap traps or glue boards in pairs. That way, if a rodent jumps over the first trap, it may land on the second. Also, if the rodent is successful at taking the bait from one snap trap, it may become less cautious when it encounters the second trap and end up getting caught.

★ Reusing snap traps that have caught rodents makes them user friendly. The rodent smell on them will dispel suspicions. For better luck don't actually set the traps for a few days, but just put bait on them to accustom the rodents to eating there.

★ Use fresh bait, and change any bait that becomes stale. Stale bait loses its smell and rodents may ignore it. Just about any food will work as bait: cheese, peanut butter, nut meats, bacon, raisins, or a hunk of bread with butter on it. Tie a solid bait like cheese to the trap trigger with a small piece of thread or string, so the rodent can't just depart with it or gently nibble it without springing the trap.

Halting Flies

You can usually reduce the fly population in your barn just by closing any windows or doors for the hours that they're in direct sunlight. Most types of flies prefer to congregate in sunny areas outside the barn, unless the odor of manure and horses attracts them indoors. Remove manure and wasted feed from stalls daily, and compost them in a covered pile.

If you need to leave the windows open, install screens. Coating the screens with insecticide can be very beneficial. A fan in the stall will also discourage flies, especially gnats and smaller insects that are not strong fliers. Other fly-control devices include fly traps, flypaper, and electronic fly zappers. A zapper attracts flies to its light and kills them on contact. A fly strip can be used in enclosed areas like feed rooms and tack rooms. A jar trap (which flies enter but cannot exit) uses attractants to lure the flies. Once a few flies die inside, the smell alone will attract more flies. Although a jar trap works well, the offensive odor may limit where you want it located. You'll need to empty or replace it periodically, as you will any type of flypaper or sticky bait.

Curbing Flies with Beneficial Wasps

Another method of fly control is to use parasitic wasps. These stingless wasps are harmless to people and animals. They spend their lives on or near manure, where

ELUSIVE RATS

If you have a rat that's too suspicious to take regular bait, try a muskrat trap covered with a handkerchief or a scrap of fabric. Rats are curious and will usually investigate a new or strange item. But you'll need to securely anchor this type of trap to something solid, or a rat with just a leg caught in it may run off with it.

the females lay eggs in the pupae of houseflies, stable flies, and several other types of flies. The wasp eggs quickly hatch and then use the dormant fly pupae as food, killing the fly before it develops.

The wasps are usually present wherever there are flies, but not in large enough numbers to control the fly population. However, you can make a significant difference by buying and releasing more wasps at the beginning of fly season. Suppliers recommend spreading them directly on and around manure in corrals and barns in early spring before flies are numerous, and then putting out more every 30 days. The number needed will depend on how many horses you have. A stable area with one or two horses will require about 5,000 wasps each month. A facility with a larger herd will need 1,000 to 2,000 wasps per horse per month. If you use this biological method of fly control, don't use pesticides, or you'll kill the wasps along with the flies.

More information on how to use these wasps is available online at www.cdpr.ca.gov/docs/ipminov/bscover.htm.

KEEPING BIRDS OUT

If birds are a problem in your barn, one way to keep them out is to install a revolving amber light (available at auto supply stores) in the loft or near the rafters. The moving light will deter them.

Controlling Flies with Nematodes

Another natural way to control flies is with nematodes, microscopic worms that live in soil. They release bacteria that are harmless to mammals but destroy flies, termites, and other pests in their larval stage. Various companies, including some listed in "Suppliers of Beneficial Organisms in North America" as well as BioControl Network (www.biocontrolnetwork.com), sell them. After receiving a shipment of nematodes, add water to activate them, and place them in manure piles, under water troughs, in stall corners, and around any other sites where flies are a problem.

BARN FIRES

Every year, dozens of horses are killed in barn fires. Most fires, however, are preventable. Should a fire occur, it's imperative to act quickly and rationally.

★　★　★

Fire-Prevention Tips

With all the material in a barn that can burn rapidly, fire is always a potential danger to a stabled horse. Don't take a chance with the well-being of your horses. Follow these safety recommendations.

★ A fire in straw bedding can burn more than a 10-foot circle (about the size of a stall) in less than three minutes. To minimize fire risk, don't store hay and bedding materials in the same building with horses. If separate storage isn't feasible, a loft is better than stall-level storage because heat and fire rise. Consequently, you'll have more time to move horses out of the lower level in the event of a loft fire.

★ Never store highly volatile, flammable materials, such as kerosene, gas, paint, fertilizers, and insect repellents in a barn. Be aware of other sources of heat that may ignite bedding, dust, or even cobwebs: matches; cigarettes; electrical appliances; fence chargers; motors; electrical fixtures and wires; batteries; sunlight through broken glass; and sparks from welding, machinery, or a farrier's work while hot shoeing.

★ Keep a barn free of cobwebs and dust. Flames can travel along cobwebs on ceilings or in rafters, spreading to other areas and possibly dropping into a stall while burning. Vacuuming is a better choice than sweeping, as it minimizes respiratory irritation to horses. You can buy or rent industrial-strength machines; if the attachment won't reach the ceiling, use PVC pipe as an extension.

Firefighting Resources

In most cases, a handheld fire extinguisher can control a wood, chemical, or electrical fire in its early stages. It's safer than water in fighting some fires; water can splash chemicals or conduct electricity and deliver a fatal shock to the person handling the hose. Mount fire extinguishers no higher than eye level in handy locations: all entrances and exits, in the middle of long aisles, in the tack room and storage areas, near the electrical panel box, and outside the barn. Service them annually.

Because many barn fires originate in wood or straw, however, a water supply is often your best suppression weapon. You can create an inexpensive sprinkler system by attaching a perforated garden hose to the ceiling along the whole length of the barn and hooking it to its own water supply. For emergency access, make sure the water source as well as the switch to any stall sprinklers is near a door.

Another alternative is to keep a hose (at least as long as the barn) on a reel mounted next to a reliable water source. Keep a click-on spray nozzle nearby if the hose is used for other purposes.

Fire Exits for Horses

When building or remodeling a barn, opt for wide aisles and plenty of exits, so you can lead a horse directly outside without having to go more than 100 feet. Make exit openings as wide as the aisle. Doors should open outward or slide completely to one side, with latches that open with one hand. If horses are not wearing them, a halter and lead shank should be hung on or beside each stall door (but out of reach of the horse).

An outdoor stall run should have its own gate to the outside in addition to the stall or barn door so that you don't have to take horses in the turnout area through a burning barn or tear down the fence to rescue them in case of fire.

Rescuing Horses from a Burning Barn

In the event you ever need to lead horses from a burning building, having an advance plan will make it easier for you to do so as quickly, quietly, and calmly as possible.

★ Sometimes a horse will freeze up. If a halter or lead shank isn't handy, use a rope, baling twine, a belt, a scarf, or anything you can place around his neck close to his head.

★ A blindfold often helps when a horse balks in a fire. Cover his eyes and nose if you have a jacket or grain sack. If he still refuses to lead, you may be able to convince him to move by pulling him sideways, pivoting around you as you press on his shoulder. You could also use a rump rope or tap his hindquarters with a whip. If other people are with you, try locking arms behind him to move him forward. Alternatively, you may be able to back him out the door if he refuses to go forward.

★ Once outside the burning barn, put the horse in a safe paddock, tie him, or have someone hold him. Never turn him loose except as a last resort; a panicked horse may run back into the barn, because it's his home and a source of security. To be safe, keep him quite a distance from the barn, where he'll be away from the heat (it can be overwhelming even 100 feet away) and able to breathe clean air with no smoke.

★ Even if a horse comes out of a burning barn seemingly uninjured, have him checked by a veterinarian. Lung damage from smoke is the biggest killer in a barn fire; it can lead to fatal pneumonia.

COMPOSTING

Daily barn cleanings can be a valuable source of fertilizer if properly composted.

★　　★　　★

Using Manure as Fertilizer

Each horse produces the equivalent of about eight or nine bags of good fertilizer annually. A ton of nitrogen-rich horse manure supplies as many or more plant nutrients as there are in most 100-pound sacks of commercial fertilizer. In addition, manure adds valuable organic matter and trace minerals to the soil.

Horse manure is different than cow manure. It has the perfect pH for roses, azaleas, and evergreens, but before using it on most plants or spreading it on fields, you should compost it (preferably with straw bedding) to get rid of excess acids, salts, and ammonia. Composting not only kills most worm eggs, fly larvae, and weed seeds, but it cuts down on the odor. Although composted manure contains slow-release nitrogen, it should still be applied sparingly; when spreading it on lawns or pastures, apply a layer no thicker than ¼ inch once a year, preferably in early spring.

Creating Compost Piles

Because a mixture of manure and bedding takes several weeks (or months) to break down in a compost pile, it's handy to have three piles: one for fresh barn cleanings, another in the process of decomposing, and a third that is well composted and ready to spread on your garden, lawn, or pasture. After you haul the oldest one away, you can start a new fresh pile. The composting area can be a concrete pad, a series of retaining walls or wood bins (each 8 feet long, 8 feet wide, and 4 feet high), a shed, several large garbage containers, or even covered piles on dry ground.

To break down at a fast rate, a pile must be at least 1 cubic yard or more in volume and contain the correct ratio of manure, bulk (straw or shavings from bedding), water, and oxygen. The proper amount of moisture is also crucial to the fermentation process. Most stall cleanings will fit the bill if they are a mix of manure and bedding material. But if stall cleanings contain too many shavings and not enough manure, the ratio of carbon to nitrogen may be too high for good compost; shavings that haven't completely decomposed will rob nitrogen from the soil, turning plants yellow when used as fertilizer.

Covering a compost pile with a lid or tarp will keep it from excessively drying in the sun or becoming too wet in the rain, which will cause the nutrients that make good fertilizer to leach out and contaminate groundwater or streams. Don't put a compost pile in a low area where moisture could collect and make it soggy; instead, locate it on high ground, away from wells or other water sources.

A properly constructed compost pile that is well maintained (turning the material periodically and monitoring moisture content and heat production) can become fertilizer in as few as 90 days, but the average time is 120 days. However, a pile that doesn't have the right ratio of materials or isn't cared for may take 6 to 12 months to become good fertilizer. You can tell when compost is ready to use: The pile is half its original size and the material is crumbly like loamy soil, with a good earthy smell.

IF YOU CAN'T COMPOST

When you don't have room for a compost pile or a pasture to spread it on, you may be able to sell or give away horse manure to home gardeners (particularly people who have rose gardens), organic farmers, or mushroom growers (the latter appreciate manure composted with straw bedding; it will not be suitable if you use wood chips or shavings, which do not produce the proper pH). If you live near a company that sells natural fertilizer, it may want your manure for its own composting process.

Tending Compost

Once the manure and old bedding is in a pile, the temperature will rise due to fermentation. Because a pile becomes hotter at the center, it must be turned and mixed periodically in order for the whole pile to break down efficiently. Turn a pile three or four times in the first two months, and then less often as the pile ages. A pile that gets too cool will break down slowly and won't kill parasites or weed seeds.

A pile that grows too hot (above 160°F/71°C) begins to kill the microbes that drive fermentation, and may actually burn itself up rather than become compost. A heap taller than 10 feet and more than 15 feet in diameter that gets too hot inside may start to slowly burn. If it becomes too moist from rain or flooding, the combustion process will speed up and the pile may burst into flame, much as wet hay sometimes spontaneously ignites.

If a pile gets too hot, make it smaller. Conversely, if it stays too cool, make it larger. If you don't want to bother turning the material, just start a new pile when the old one gets big, and allow a year for each pile to compost. Cover the piles with plastic or dirt to prevent them from drying out too much and to keep nutrients from being leached away by rain or snowmelt.

CHAPTER THREE

Fencing

Making sure that fences and gates are in good repair is an important part of keeping horses. Here are some words of advice on building and maintaining safe, strong enclosures.

3

FENCE CONSTRUCTION

The extra time and effort you spend installing proper fencing for your horses will pay off in the end with enclosures that are both safe and lasting.

★　　★　　★

Pipe Fence

Pipe rails make durable fencing and are often used for stallion pens or boundary fence along roads or other areas where it is important to have a secure fence. It can be dangerous, however, if the spacing allows a horse to catch his head or a leg between the rails. A horse putting his head through the rails in an attempt to chew on the next rail down could catch his mouth on the lower pipe. As he struggles to free himself, the upper pipe pressing into his windpipe could suffocate him. To avoid this danger, make sure the rails are spaced close enough together, or attach three or four strands of cable between the top and bottom pipes.

Wood Fence

Create a smooth fence line by attaching the rails and boards or other siding to the inside of the posts, even though they look neater nailed to the outside. That way, a horse dashing along the fence won't run into them or knock them off by rubbing, leaning on, or kicking the fence.

Wire-Mesh Fence

To keep the wire from sagging, brace it well at the corners, and top it with a pole, pipe, board, or electric wire to keep horses from mashing it

FENCE HEIGHT

For safety's sake, a fence should be as tall as a horse's withers (5 feet high for a 15-hand horse). Corrals, small pens, and stallion pens should be even taller, preferably at the horse's eye level when he holds his head up in a normal position (usually at least 6 feet high).

down. Some types of mesh stay tighter and last longer if you put a pipe or pole at the bottom as well as at the top. Install the mesh on the inside of the posts so that horses running into it or rubbing on it won't push it loose or pop the staples out.

Woven mesh is a better choice than welded mesh: It's stronger, safer (if a weld breaks, a piece of wire may stick out and snag a horse), and less vulnerable to rusting. Galvanized wire, which is dipped in hot zinc oxide, doesn't rust as readily as untreated wire.

Plastic Grid Fence for Foals

The soft plastic mesh used around construction sites makes safe, foolproof fencing for young foals; they can't get a foot through it, and it won't hurt them if they run into it. Used mesh is sometimes available inexpensively. To keep the mare from leaning over it or pushing it, install electric fence wire or tape along the top or just above it.

Chain-Link Fence

This type of fence needs a pipe at the top and the bottom to brace it, and it should always be high enough that a horse can't get his head over it. The height is important because, even with a top pipe, the ridges at the top of the fence can cut a horse's throat or jaw. Another drawback to chain-link is that a horse can catch his shoe in it if he paws the fence.

POSITIONING ELECTRIC FENCE GATE HANDLES

Be sure a gate handle disconnects from the side of the gate nearest the charger so that the gate wire is dead when unhooked. This will allow you to test the entire fence leading from the charger to the gate; you simply touch the gate handle to its hookup and see if you have a spark. Also, the unhooked, open gate wire won't shock you or a horse, and there's no danger of it sparking a fire in dry grass or old manure if you put the wire on the ground or throw it over a fence out of the way.

Electric Fence

Hot wires are a quick and inexpensive way to contain horses or make existing fences safer and longer lasting, but they need constant maintenance. Keep grass and branches trimmed away from the fence when they are growing rapidly, lest they smother the wires and short out the fence.

Make sure the insulators are smooth, with no sharp protrusions. If you're using metal T posts, cap the posts to form a smooth, round top (although some horses will fiddle with caps and take them off). Another option is to use the smooth-top flexible posts designed for electric fencing.

Test the wire often to be certain it's working. Most types will break or stretch (possibly touching the ground or a metal post or fence) and short out if horses or wildlife run into it. If you have a fence charger with a blinking light that indicates electrical current is running through the wire, glance at it at least once a day while doing your chores to make sure the light is still blinking (a sign that the fence is working).

Another method is to check the gate handles at strategic locations, to find out if the fence is working up to that part. When you unhook a handle, the light on the charger should start blinking again if the short is beyond the gate, and you'll know which portion of fence to look at more closely to find the problem. If you have several handles, you can quickly pinpoint the area of a short and thus save time walking the fence line. But if you have a long stretch of fence with no gate handles, consider adding some: They'll save you a lot of time over the years.

Choosing a Fence Charger

Because horses are sensitive to electricity, it doesn't take much current to keep them from touching a fence. Use a type of fence charger with a pulsating current breaker rather than a steady charge. Some electric fences are too "hot" for horses (because they're so well grounded, with four feet on the ground) and can give them a serious shock, or even electrocute them under certain situations.

You can select a charger that operates on electricity, batteries, or (in a sunny climate) solar power. An electric or battery-operated fence charger should be protected from weather, in an easily accessible location where you can turn the charger on or off. A solar charger, on the other hand, can be hung anywhere along the fence line in a south-facing spot to get the most sunlight. It doesn't need protection from the weather.

A battery-operated charger will need a new battery every two months. If you use a dry-cell battery, keep the box away from concrete surfaces, which may drain its charge. Insulate the box by placing a rubber mat underneath and behind it.

With a charger that's in a barn or other building, place a gate handle near the start of the fence so you can unhook it quickly from outside when you want to turn the fence off to repair a section of it, or to untangle a horse that's caught without being shocked by a hot wire.

PROTECTING A CHARGER THAT'S OUTDOORS

When there's no building to house the fence charger (say you're using a battery charger to electrify a pasture fence far from your barn), you can make a hinged plywood box to enclose it. To protect the box, coat the plywood with a waterproofing material and, if you like, cover the top with galvanized metal roofing. Using an acrylic plastic panel (screwed and caulked in place) for one of the sides will allow you to see whether the charger's light is working.

Grounding the Charger

A fence charger needs a proper ground in order to work. An electric fence is grounded at its beginning, and the electrical current runs along the fence wire, trying to return to the ground to complete the circuit. As long as the fence is insulated and there's nothing along it that can carry current to the ground, each electrical pulse will continue to the end of the wire. A short circuit, or a shorter pathway to the ground, occurs when the wire touches something that conducts electricity (such as a steel post, wet wood, green grass, or a horse's body).

The ground for your charger can be a steel rod or water pipe driven into the ground and connected to the charger with sturdy insulated wire. The end of the wire should be stripped bare and wound several times around the rod or pipe to ensure good contact. The wire going from the charger box to the fence should be insulated, so it won't shock anyone or pose a fire hazard. Don't use nails, metal clips, or staples to support the wire on its way to the fence, because the points of attachment can become worn over time, creating a short that is hard to locate. Use insulators and secure the wire just as you would to a bare fence wire, so that if the insulation around the wire ever cracks or frays, it won't create a short.

Making Electric Fence Wire Visible

A traditional electric fence wire isn't very noticeable. To keep a horse from running into the wire before he sees it (especially when he's put into the enclosure for the first time), flag the wire with bright strips of cloth or fluorescent orange surveyor tape tied every few feet. The newer, brightly colored electric tapes (with small wires woven into the plastic) are more visible, as well as easy to install or repair: You can fix a break in the fence just by tying the broken pieces back together. One disadvantage of some of these tapes is that they don't last as long as wire; sunlight damages them and after a few years they become brittle or the plastic disintegrates.

If you use the broader electric tapes, install the posts or supports closer together. The tapes tend to blow in the wind (and may touch something and short out), and sag to the ground under heavy snow.

GATES & LATCHES

Gates should be safe and strong, easy to open and close, and have latches a horse can't open.

★   ★   ★

Keeping the Gate off the Ground

A heavy gate will put less stress on its hinge post when the other end rests on a block of wood at the base of the post where it latches. If a gate rests on the ground, it can be hard to open and close, and the bottom pole or bar may rust or rot if the area is wet or muddy.

Setting a Gatepost

A gatepost must be sturdy enough and set deep enough to never shift, and properly treated so it won't rot off. A railroad tie works well; it is large and sturdy, and pressure treated on all surfaces. To hold a heavy gate, the post should be set at least 3 to 4 feet in the ground and well tamped. If you have wet soils, mix some gravel with a little dirt (you can buy some gravel if you don't have any on your place). Tightly tamp the mixture into the post hole a little at a time until it's solidly filled.

To fix a leaning gatepost (that you don't want to take out and reset), push it back into a vertical position with a tractor or vehicle, and fill the gaps with soil and gravel, tamping well. Alternatively, drive a wedge of pressure-treated wood into the space. In some cases, you may decide to stub the leaning post and set a new post next to it on the outside of the pen, where a horse won't run into it. You can bolt or strap the two posts together with several loops of smooth wire pulled tightly and stapled.

Some soils are hard to keep posts in, and in cold climates, frost may heave posts out of the ground. In those cases, you must reset the post, first enlarging the hole to accommodate a longer, deeper post with a T bottom, an anchor you create by securely bolting a crosspiece on the end. Another way to keep a gatepost from shifting is to cap the gate. This requires setting very tall posts deep in the ground and attaching a large pole or beam in place across the top of both gateposts, well above the horse's head. The two posts brace one another and minimize shifting.

Bracing Gateposts

In addition to being deep and well tamped, a gatepost must be well braced to open and close properly. On a wood fence, the rails between the gatepost and the post next to it act as a brace. On a wire or mesh fence, you can create a brace by nailing a sturdy pole between the gatepost and the next post (on the outside of the netting, so the horses won't chew it) midway up the posts or near the top. For a very secure brace, attach a horizontal pole between the tops of these two posts and a diagonal pole on the next section, running from the top of the first brace post to the bottom of the second one.

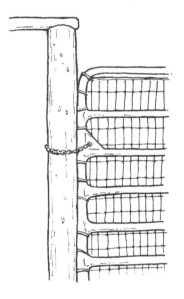

Mesh netting wired to a gate

Making a Gate Safer

Like fencing, gates should always be smooth to prevent injury to horses. Allow a minimal gap between the gate and each gatepost, so a horse or foal can't catch his foot or head in it. On a metal gate, put the upper hinge as high on the gate as possible. If the spacing on the bars or poles of a metal gate is too large, you can add more poles or weld more bars between the open spaces. Another option is to use mesh netting (V-mesh wire weave fencing is safer for foals) cut to fit and securely wired in place.

Insulating Electric Wire That Spans Gates

Strategically placed electric fence wire or tapes can protect wood fences from being chewed, leaned on, and rubbed against, by teaching horses to stay away from the fence and not reach through it. When the electric wire must span a metal gate, however, it can cause problems if it's too close to the metal. The wire could touch the gate (for example, when a strong wind blows it) and short out the fence. It could also electrify the gate and shock you when you touch it.

An easy way to prevent problems is to thread the electric wire through a length of old garden hose to insulate it where it spans the metal gate. For this purpose, use a piece of fairly stiff, smooth wire about the length of the hose (it will push through the hose more easily than a smaller, more flexible wire). Attach an electric fence handle at one end for disconnecting the electric wire when you open the gate. The handle combined with the hose will keep the wire from shorting out on the gate.

A piece of old garden hose over the wire also works very well wherever wire goes through a wall, next to a haystack, or any other place that you don't want it to touch and accidentally short out, shock someone, or cause a fire.

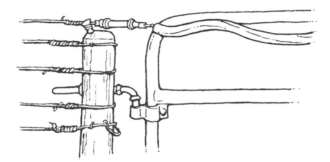

Use a garden hose to insulate electric wire.

BYPASSING A GATE WITH ELECTRIC WIRE

If you don't want to deal with a hot wire every time you go through a gate, you can run it high enough above the gate that a person, horse, or vehicle will never touch it. Use tall, lightweight poles at either end of the gate to gain the needed height.

Making a Gate Easier to Open

A wood or metal gate becomes a heavy burden to open and close if it sags and drags on the ground. To solve this problem, attach a small wheel (for example, from a wheelbarrow) to the moving end of the gate to support the weight and prevent the gate from sagging any lower. You can simply bolt the upright portion of the tire holder (the piece of metal that comes down each side of the tire to hold its small axle) to a wood gate or wire it securely to a metal gate.

Alternatively, you can securely wire an old wheel or small tire with any kind of long axle to the gate. Attach the axle at both ends (close to the wheel and the far end) to the bottom rail of the gate with stiff, strong wire to keep the weight of the gate from altering the angle of the wheel or causing it to bind. A wheel is also useful for opening and closing a heavy, wood-pole panel in any opening where you don't have a sturdy gatepost to hang a gate.

Bolted wheel

Wired wheel

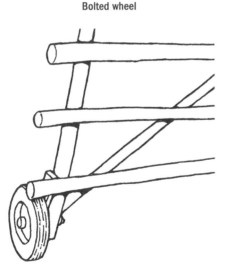 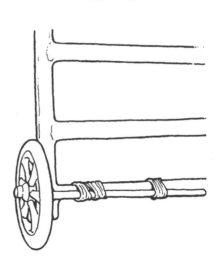

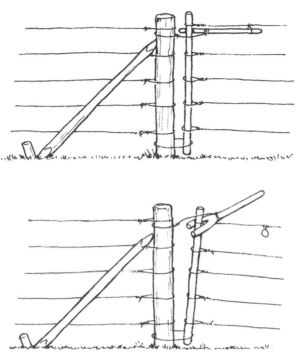

Straight "jimmy stick" gate closer

Forked stick gate closer

Making a Gate Easier to Latch

A gate made of wire rather than wood or metal pipes or rails is often used in pastures that have livestock as well as horses. To close it, slip the top and bottom of the gatepost into wire loops attached to the top and bottom of the adjacent fence post.

If the gate becomes difficult to close, use a "jimmy stick" (a 2- to 3-foot-long pole with a diameter of 1½ to 2 inches and wired to the main gatepost) as a pry bar to move the wire gate close enough to the gatepost to fit the top wire loop over it. To be effective, the jimmy stick's attachment wire needs just enough slack to enable a person to use the pole for leverage. Wrap the wire around the stick and staple it about 5 or 6 inches from one end. To use the jimmy stick, press the short end against the end post of the gate, then push the long end toward the gate top to pull the gate closer to the gatepost.

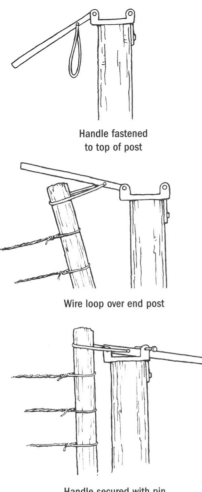

Handle fastened
to top of post

Wire loop over end post

Handle secured with pin

A slightly different version uses a stout forked stick about 3 feet long, wired to the gatepost. You place the forked part over the gate top, push the long end parallel to the top gate wire, and tuck the pole end into a wire loop located on that wire.

Another way to make the gate easier to close is to use a metal handle with a wire loop attached. Fasten the handle to a flanged platform bolted to the top of the post. When open, the handle gives you an extra 12 to 18 inches of reach for maneuvering the end post of the gate into the wire loop. Once the loop is over the post, you can just push the handle up and over, automatically bringing the gate closer to the gatepost. You can secure the closed metal handle with a pin in the raised metal tab; this will keep the handle from popping up if a cow or horse rubs on it.

Closing a Gate Automatically

You can make a small, lightweight gate automatically close behind you by adding a counterweight. Securely attach a metal ring to the gatepost near the top of the gate and thread a thin, smooth rope through it. Staple one end of the rope to the far end of the gate, then run the other end down the gatepost and suspend a weight from it. The weight tied to the rope must be just right: too heavy and the gate will slam shut; too light and the gate won't close properly. A small bucket of sand is ideal, because you can add or remove sand until the weight is correct. When the gate is opened, the weight will pull it closed after you let go. Install a self-locking latch and you won't have to worry about fastening the gate when you have your hands full, for example, leading horses or carrying feed.

Horseshoe Gate Latch

A horseshoe welded or securely wired to the proper length of chain can serve as a latch on a pole gate, provided the horseshoe isn't too large to fit securely onto one of the gate's horizontal poles. At the height you want the gate latch, nail or staple the free end of the chain to the gatepost, making sure it won't come lose. When the gate is shut, slip the horseshoe over one of the gate poles.

Dealing with Frozen Latches

To keep a snap latch from routinely freezing in cold weather, always turn it so the moving part is underneath rather than on top; that way, moisture from rain or snow will drain off. It may still freeze up in a blizzard, however. If so, pour hot water on it or blow your warm breath on it a few times. Where cold mornings are common, carry a tube of de-icer (for car door locks) or de-icer spray (for windshields) in your coat pocket when you do chores. Another way to try to prevent snaps from freezing is to put vegetable oil, mineral oil, or any kind of lubricating oil on the inside or on any moving part of the snap to seal it from moisture.

HINTS *for*
MAKING LATCHES FOOLPROOF

An ordinary latch may not keep your horse inside a pen or stall. Here are some modifications you can make.

Latch Protector. If a horse plays with a latch, attach a small, rectangular piece of metal above the latch to cover it. Because this latch requires one hand to lift the cover and another to undo the latch, the horse can nibble all he wants to no avail.

Two Latches. On some gates you may want two latches: a secure chain that horses can never unhook, and a quick and easy latch for the times you are going through the gate repeatedly and don't have both hands free for opening and snapping the chain. A handy latch for those times is one that secures the gate automatically when it swings shut against the gatepost.

Stall Door Latch Tricks. One way to keep a horse from opening his stall door latch is to install another one partway down the door on the outside, well out of his reach. Another trick is to tie a piece of baling twine through the latch, or thread a piece of wire through it and twist the two ends together to hold it in place.

Seat Belt Latch. If a horse is good at opening latches and letting himself out, you can use an old car seat belt (obtained at an auto salvage yard) as a safety latch. Bolt the webbing next to the insert end of the latch to the gatepost, and bolt the release-button end to the gate (or stall door), first drilling holes in the wood where needed. Once the seat belt is snapped together, a horse usually can't press the button to open it. This type of latch is durable and works well even in cold weather, unlike chain snaps, which tend to freeze stuck when moist.

★

MAINTENANCE & REPAIRS

Prevention can save time and money, particularly if you find ways to keep horses from damaging a fence, and keep up with small repairs that head off big problems.

★   ★   ★

Fence Maintenance Tips

No fence lasts forever, but a little care now and then can extend its life for many years. Do a thorough check several times a year, walking all the fences. Write down things that need to be fixed, and where. This will also help you figure out what materials and tools you'll need for the repairs.

★ Assess each post to see if it's loose, out of line, or rotting, so you'll know if it needs to be replaced or just retamped. Check each wood post at ground level with a pocket knife to see if it's deteriorating (it'll slide right into rotted wood). See if any wood panels are badly chewed or starting to rot.

★ Check wire mesh for places that sag and need tightening, bracing, or a pole along the top or bottom. Look for breaks and exposed sharp wires or rusted bottom wires that are about to break. If the mesh goes clear to the ground, the bottom wires will eventually rust, especially if they become covered with horse manure. Look at metal posts to see if they are bent or rusted.

★ Examine fences made from synthetic materials for discoloration and cracking. A PVC fence may crack in cold weather. In time it will deteriorate and break more easily. In some cases if the posts move the planks may slide out.

★ Check rubber or nylon-strip fencing for sagging or disrepair. Trim off any frayed edges or unraveled pieces with a sharp knife or scissors and heatseal the rough edges with a candle or lighter.

HOMEMADE WOOD PRESERVATIVE

To keep wood fences from weathering and deteriorating, periodically apply a good wood preservative (one that isn't toxic to horses) as a protection from moisture. How often will depend on your climate. In drier regions, it may be only once a year; in rainy areas, several times.

Good commercial products are available, but you can make your own by stirring 2 ounces of shaved paraffin or candle wax in ½ gallon of kerosene until it dissolves, then adding ½ gallon of boiled linseed oil. Brush or spray (with a garden sprayer) the mixture onto the fencing until the wood won't absorb any more.

Halting Wood Chewing

You can buy antichew products to keep horses from chewing wood. Another option is to use foul-tasting home remedies that are safe for horses.

★ Rub a bar of soap over favorite chew spots, or pour a little dish soap on the wood.

★ Most horses don't like the taste of hot peppers, which can be safely used in a homemade mix for brushing on wood (and also on a mare's mane and tail if her foal likes to chew on them). Combine Tabasco sauce, cayenne pepper, and a little water. If that's not strong enough, try adding juice drained from a jar of hot peppers to the mixture and let it set (in a covered container) for a couple of days.

★ Mix together ½ cup of black pepper, 1 cup of vinegar, and a few finely chopped garlic cloves in a saucepan. Cover and boil the mixture for five minutes, until it is the consistency of thin paste that you can brush on wood.

Reinforcing Wood Fences with Wire

When confined in small pens, many horses can ruin a good fence in short time by chewing on it. Wood preservatives, old motor oil, and other foul-tasting products used by horsemen will deter some chewers, but some of these chemicals can be toxic. Old motor oil, for instance, contains lead.

A safer way to protect fences is to cover the wood with small-mesh chicken wire, cut to fit around poles, posts, or boards. Horses find the chicken wire unpleasant on their teeth and leave the wood alone. A roll of chicken wire will go a long way when you cut it into strips. Staple the mesh at close intervals to avoid sharp protrusions and loose edges. Use staples large enough to hold securely and never pull out. When you weigh the costs of the wire, staples, and your labor, it will be a lot cheaper than having to replace chewed-up poles, posts or boards. If you want to, you can still periodically brush a nontoxic wood preservative, such as log oil, right over the chicken wire.

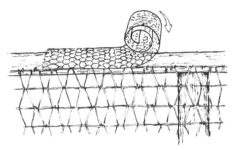

Applying strips of chicken wire

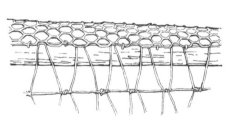

Chicken wire stapled in place

A similar way to protect posts, poles, or boards from chewing is to wrap wire around them. Baling wire works well, but very few farmers use it anymore. Instead, you can buy or recycle electric fence wire. Staple one end to the fence, then wrap the wire tightly around the pole or post, with about 1 inch of spacing between loops. If the wire is in a roll or coil, you can easily pass it around and around the pole. Staple it periodically to keep the wire from coming loose or moving along the pole or post.

STAPLING TIP

When stapling wire or wire mesh to wood posts as primary fencing (rather than using it to cover an existing fence), don't drive the staples in completely. Leave a tiny bit of space between the wire and the staple so the fence will have some give and be less apt to break if an animal hits it. That extra space will make repairs easier: If you ever need to retighten the fence, you can use a fence stretcher and the wire beneath the staples will be able to move instead of being bound solidly to each post.

Keeping Others Out

A good way to keep dogs, wildlife, and children out of horse pens and pastures (and prevent horses from sticking their heads through the rails) is to add wire mesh to the outside of the existing fence. Use a small mesh such as 2 x 4-inch, no-climb netting, diamond-mesh fencing, or V-mesh wire, because an adult horse can't put his foot through it. You can purchase heavy-gauge mesh in 100-foot rolls; it comes in various heights, so you should be able to find a match for your fence.

Use smooth wire to fasten the mesh to the fence at one end. As one person rolls out the mesh, stretching it as tightly as possible, another person can staple it to the fence rails and posts (a staple gun can make this job faster). If it's a long distance, after you get the mesh rolled out, use a fence stretcher come-along pulley on the end to stretch it fully before you start stapling.

When covering a pipe fence, attach the wire mesh by twisting smooth wire around the pipe and using pliers to tighten the connection completely and crimp the ends so they won't poke a horse. It's best if the wire ends are on the opposite side of where the horses will be. Another option is to use plastic electrical grip ties to secure the mesh to the pipe.

TEACHING A HORSE BOUNDARIES

Before turning any horse loose in a new paddock or pasture, acquaint him with the boundary fence, especially if it's not highly visible, by leading or riding him around the inside. Unless a wire fence has an obvious top pole or other solid portion, put flagging on it.

Emergency Repairs

Sometimes you'll encounter a problem that can't wait until you return with materials or tools. For example, wildlife may tear down a fence or a tree may fall on it. Unless you make a temporary emergency repair, your horses could get out or be injured by the damaged fence.

An old cowboy trick for tightening loose wires between two posts when you have no tools or materials on hand to fix it is to weave several pieces of wood (willow branches, dead sagebrush, or small-diameter tree limbs) through the wires, with the ends alternately placed in front and in back of each wire. The lacing effect (like basket weaving) takes all the slack out of it.

Quick Fix for a Missing Insulator

If an insulator on your electric fence is missing or damaged, here's a temporary fix until you can replace it: Join the electric wire or tape back to the post with baling twine, first wrapping the post (or the broken insulator) with twine to keep the wire or tape from coming in contact with the post itself. Baling twine is good insulating material and will keep the wire from shorting out.

Mending a Fence Board

A horse can quickly chew a board all the way through. Once he starts on a certain spot, he'll return to it again and again because it's easier to sink his teeth into it or even tear hunks and splinters out of it. If the board is damaged in just one area (whether a horse or something else is responsible), you can repair it instead of replacing it.

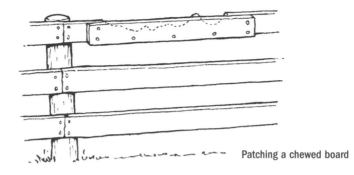

Patching a chewed board

Cut a length of board about a foot longer than the chewed or cracked area, place the patch over the weakened spot, and screw it into the solid wood at each end (a battery-powered drill will make the job easier). Either have someone hold the piece in place while you screw it in, or use clamps if you are working alone. Aim the screws at a slight angle so they won't go clear through the wood, to avoid any sharp protruding tips that could injure a horse. Use at least four screws to secure the top and bottom of each end of the patch. If it's longer than a foot or so, use more screws along the bottom where the original board is still intact. Apply a bad-tasting, nontoxic deterrent if your horse is likely to chew the spot again.

Handy Pole or Board Holder

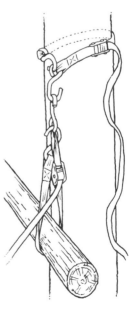

When building or repairing a pole or board fence by yourself, or when putting up poles or boards for any type of construction, it can be difficult to keep the far end of the wood in place as you nail the other end. An easy way to secure the far end until you can nail it is with motorcycle tie-down straps. These are adjustable nylon straps with a hook on each end and an adjustable tightener with a thumb release. Loop one motorcycle strap around the post (above where you want the pole or board), tightening it so it can't slip. Loop another strap around the pole or board and then hook the two straps together to hold the board at the proper height while you nail the other end.

Tightening Wire without a Fence Stretcher

Sagging or broken wires can be dangerous if a horse tries to reach through or over the fence, or if he puts a foot through it. Even when you don't have a fence stretcher or come-along pulley, you can still splice and tighten broken or sagging wire with just a carpenter's hammer.

First you may need to add a short piece of wire (1 to 2 feet long) for the splice. Make a loop in one end of the broken fence wire, twisting it back around itself to hold, then run an end of the short wire through the loop and twist it back around itself, as well. Now you're ready to make the splice. Here's how to do it.

1. Make a loop in one end of the wire and pull the other end through it.

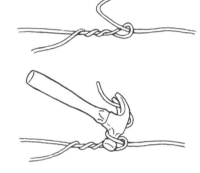

2. Anchor the loose end between the hammer claws.

3. Twist the hammer so the wire wraps around it.

4. Keep twisting until the wire is as tight as you want it. Bend the wire back toward itself, then untwist the hammer. The tight crimp will hold the wire tight.

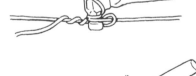

5. Finish the splice by wrapping the end of the wire tightly around itself.

Latching a Sagging Gate

A heavy-duty metal gate is much stronger and safer than an aluminum gate, and longer lasting than a wood gate, especially in places where it is opened and shut frequently. It can be frustrating, however, if it starts to sag, particularly if it has a lever-type latch with metal prongs that fit into a slot in the gatepost. As little as an inch of sag can make it necessary to lift the gate in order to slip the latch into place.

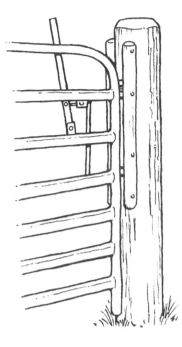

An easy way to permanently fix this problem is to nail two short poles on either side of the latch area to enclose the metal latch prongs wherever they happen to hit the gatepost. As long as you can move the latch enough to flip the metal keeper down (to hold the lever closed), the gate will stay shut. If necessary, you can use a hatchet to whittle out a slot in the gatepost so the latch will sink to full depth where it lines up on the post. Then it won't matter if the gate eventually moves as much as several inches out of alignment.

CHAPTER FOUR

Tack

The equipment you use on horses must fit properly to avoid discomfort for both the horse and the rider. In order for it to last and stay safe, you need to take care of it. Here you'll find practical advice on using and caring for various types of tack.

LEATHER

Even though some types of horse gear are made of synthetic materials, leather is still the standard, and needs certain elements of care. This applies to your tack as well as apparel, such as chaps and chinks.

★　　★　　★

Caring for New Leather

Most new leather has already been oiled enough and does not need more. Too much oil can be just as bad as too little, causing leather fibers to break down. Still, a piece of leather tack is never cleaner than when it's new, so this is your opportunity to seal the leather, before you get any dirt or sweat on it. Usually the best thing to put on new leather, especially the back of stirrup leathers on a Western saddle, is a wax-based product, such as R.M. Williams Saddle Dressing or Ray Holes Saddle Butter. Apply the saddle dressing or butter in warm conditions and buff off any waxy or sticky residue to keep dust from sticking to the leather.

Cleaning Leather

Dirt and sweat can damage leather, making it hard, dry, and brittle. Proper and frequent cleaning and conditioning will keep it more supple and less apt to deteriorate. Use saddle soap and warm water to clean a saddle or bridle. Though many people think water and leather is a bad combination, that's only for extended periods of time, for example, when you're riding in the rain. Then the leather needs to be properly dried and reconditioned. Washing with warm water is just a topical cleaning that actually removes salt residue from sweat and opens the pores of the leather to float out dirt.

A nylon bristle brush is a good tool for scrubbing and lathering, and an old toothbrush works well for cleaning around buckles and other hard-to-reach places. Use a sponge to rinse off the soapy water, squeezing it out often, and then to rub on more saddle soap as you continue cleaning. Occasionally take the tack apart so you can clean all the inaccessible areas.

After you wash the leather, let it dry at room temperature, not in high heat. Follow up with a leather conditioner to restore the oils. If the leather is really dried out, apply a high grade of oil to replace what time and washing have removed. Use it sparingly, though: Overoiling adds unnecessary weight to the saddle and may also bleed out of the leather during hot weather.

Applying Oil

Remember, leather was once an animal's skin. Because it's composed of fibers held together by protein bonds, it's "healthiest" when it contains a certain amount of natural oils. As the oils evaporate over time, or if the leather is exposed to too much sun and heat, the protein bonds break down and the fibers pull apart, weakening the leather. Replace the oils with a good leather conditioner after each thorough cleaning.

A handy way to apply oil is with a $5/8$-inch artist's paintbrush. This makes for less mess and waste than using a sponge or rag because the brush can be dipped into any leather product container without having to tip it and risk spilling the oil. It's also a nice size for oiling narrow pieces of leather, such as bridle parts and all of the creases and crevices on your saddle.

LEATHER

USING LEATHER WIPES

You can purchase handy leather wipes that contain a cleaner and/or conditioner. They're too pricey to use on a regular basis and no substitute for proper cleaning and conditioning, but convenient in a pinch. You may want to try them when you don't have time to clean your tack thoroughly, such as at a show, when you're traveling, or to give your tack a quick polish after a long trail ride.

KEEPING THE TACK ROOM DRY

In a wet climate, prevent a tack room from becoming too damp by placing a plastic bowl filled with moisture-attracting chemical chips (available at a grocery or drugstore) near your tack. Also save the little moisture-halting cubes that come in pill bottles and put them in an open bowl or hang them in string-tied bags you fashion from gauze.

Make-Do Leather Conditioners

If you need to recondition leather but don't have access to saddle soap or oil, use a thin coat of petroleum jelly to protect it and keep it flexible until you can condition it properly. First, make sure the leather is clean and as dry as possible. Wipe off any excess jelly with a clean rag.

Mold & Mildew

Leather tack stored in a warm, moist environment may start to mold or mildew, which will eventually damage and destroy it unless the mold is removed. An easy way to kill these organisms is to wash the leather with a very weak bleach solution (1 tablespoon of bleach mixed with ½ gallon of water), then dry it thoroughly. If the leather is stiff and hard, apply saddle soap or leather conditioner after washing it. Once mold spores get into leather, they'll grow again whenever conditions are favorable, so keep the leather in good shape.

To prevent mildew in wet climates, rub on saddle wax to seal the leather and create a moisture barrier so bacteria and mold spores won't gain access. It'll also minimize damage from sweat and rain. Apply the wax liberally and in warm conditions so that the leather will absorb it; otherwise, it will just cake on the surface or in the cracks of a carved saddle. You can use a hair drier to help work the wax in faster. To finish, buff the leather with a soft cloth so the surface is smooth and not sticky.

Restoring Rain-Soaked Leather

Unless you regularly clean and oil your tack, it will become stiff and hard when it dries after being in the rain. Well-oiled leather repels water and comes through in good shape if you wipe it dry and apply some saddle soap after you return home from a rainy ride. If the leather has not had much care and is somewhat dried out already, follow up by applying a conditioner to seal in some moisture as the leather dries, to keep it from cracking.

Restoring Old, Dry Leather

Hot water works well for cleaning new leather, but it shrinks old, dry leather, much as hot water shrinks wool. Clean old leather with luke-warm water, then condition it. If the leather is very dry, use a thin, high-quality oil that will easily penetrate. Some people use olive oil or peanut oil, but neat's-foot oil is less expensive, more available, and works just as well, as does glove oil (used for keeping leather gloves soft and pliable).

You'll be able to tell how much oil to use. Old, dry leather will drink it in, especially if the leather or oil is at room temperature or warmed by the sun. Oil should feel warm to the touch when you apply it. If necessary, heat it on a stove in a container floated in a pan of water. Don't heat the oil in a pan directly on the burner, because it may become too hot.

After you oil the tack, let it sit for a few hours or all day. If it still feels brittle, stiff, and dry after that, apply more oil. Once pliable, the leather is oiled enough. Use saddle butter or dressing (or any good wax-based product) to seal and protect the leather.

Alternatives to Leather

For centuries horse equipment has been made from leather, which is long lasting if given good care, but modern technology has produced various types of nylon to replace rope and some traditional leather gear. Nylon can be more advantageous than leather, as it needs no oiling and can be easily cleaned with soap and water or thrown in the washing machine. But after a while, nylon tends to age and fray around the edges, especially with lots of use, exposure to sunlight and weather, and many washings.

Today a lot of tack (English and Western) is made from biothane, a nylon webbing covered with a tough material that won't fray and looks and feels like plastic. Gear made from biothane is strong, durable, and lightweight, making it popular with endurance riders. It doesn't stiffen in cold weather or dry out and become brittle in hot or arid conditions, and almost any type of tack can be made from it, including harnesses, breast collars, and reins. However, reins made from biothane should incorporate a section of regular nylon or leather where you hold them, because it's hard to keep a good grip on the slick biothane.

To clean dirty or sweaty biothane, wipe it with a wet rag, dunk it in a bucket of soapy water, or toss it in the washing machine. Rinse thoroughly to get rid of any soapy residue, which can irritate a horse's skin.

ADVANTAGE OF SYNTHETIC MATERIALS

Tack made from synthetics, such as nylon or cordura, is very pliable but has little stretch or give and so will keep its shape once you get it adjusted properly. The equipment will actually conform to your horse's contours more quickly than leather, because it's thinner.

Caring for Leather Chaps

Chrome-tanned garment leather (the leather is tanned with chromium salts) is often used for show chaps or chinks. It's thinner and softer than other kinds of leather and, unlike saddle leather, it stays soft when it dries out. Chrome-tanned leather can be washed, and some people actually put it in the washing machine, but it's not wise to do so very often. A complete soaking in water can be hard on the leather. Like silk, leather is a natural fiber and more delicate than most kinds of fabric. But if you decide to launder garment-leather chaps, test a small area to make sure that washing won't alter the dye colors, and that two-tone garments are colorfast. Don't use detergent; although it'll remove dirt, it won't replace the natural lubricating oils in the leather, so your chaps will dry stiff and they'll weaken. Use a special leather care product for chrome-tanned leather; it cleans and reconditions the leather at the same time.

Brush off the surface dirt and close all the zippers so they won't scratch the leather or the washing machine. Turn smooth leather chaps inside out to protect the outer surface from abrasion. Add the cleaning product while the washer is filling, then put the leather dressing in the rinse cycle dispenser or sponge it onto the damp leather when you remove it from the machine. Unzip the damp chaps and spread them out to dry in a place protected from sunlight or direct heat, so the leather won't shrink.

Often a pair of chaps or chinks (especially those designed for work rather than show) have a vegetable-tanned yoke, belt, or side panel made from skirting leather or strap leather. Never machine wash these garments; instead, clean them the same way you clean your saddle. Use glycerin or saddle soap with a little warm water, and scrub the dirty leather with a brush if needed. Rinse the chaps and spread them out to dry. Once dry, give them a good coating of chap wax or some other leather protection product that contains a lot of beeswax to provide some waterproofing. You can oil any areas that are vegetable tanned, but never use oil on the softer chrome-tanned leather. It's so porous that it will absorb too much oil, darkening the leather and making it stretch. Some of the oil may even come through on your legs.

FITTING CHAPS

If you wear chaps or chinks, whether in the show ring or to protect your legs when riding in brush or bad weather, make sure the style you choose fits properly. You want the leather to come below the knee on the inside of your legs. If the leather ends above or just at your knee, it will rub your leg raw if you ride a lot.

LEATHER

Taking Care of Horse Boots

Horse boots will last longer and look better if you keep them clean.

★ Like all leather tack, leather leg boots should be cared for with saddle soap and conditioners.

★ Neoprene boots can be washed in soap and water and allowed to dry.

★ Rubber bell boots just need rinsing with water to remove any mud or sand that may work its way inside them and chafe the horse.

★ Nylon shipping boots, or any boots containing nylon, fleece, or Velcro but no leather can be safely cleaned in a washing machine.

SADDLES & BRIDLES

Saddles and bridles often represent a major investment or have sentimental value. Good care and proper use can help them last a lifetime and remain comfortable for you and your horse.

★　　★　　★

Saddle-Fitting Tip

The first time you put a brand-new Western saddle on your horse, it may not settle completely into position as it will after it's been used awhile. In fact, you may think it doesn't fit the horse because it's perched a little too high. Whether the saddle is handmade with a new fleece lining or a factory model with a polyester lining, it needs a few hours of the rider's weight on it to compact and settle properly. Be sure to take this into consideration.

CONDITIONER VERSUS POWDER

Some people sprinkle talcum powder between the leather pieces on a new saddle to stop the squeak (if you tip the saddle over you'll have better access). Talc may create a frictionless surface, but it does nothing to condition the leather — and it'll fall out eventually. A saddle dressing product works better.

Stopping Saddle Squeak

New leather will squeak if there is too much friction when one part moves against another. To remedy the problem, treat the back of fenders and stirrup leathers with saddle butter or dressing to smooth the leather pieces so that they slide more readily.

First loosen the stirrup leathers and fenders so you can remove the stirrups and gain access to areas that rub together. Also lift up the side jockeys a little and apply conditioner on the back, to further ease friction. Any part of the rigging below the bottom of the skirts benefits from conditioner, because it protects the leather from horse sweat.

LOOSENING THE SADDLE SKIRTS

Some saddles have skirts that sit tight against the bottom of the bars. You won't be able to put the stirrup leather back in place unless you loosen the top of the skirts by removing the screw that holds it. Turn the saddle over and dig around in the fleece (or other lining) with your fingers to find the screw or nail head and loosen it.

Removing Stirrup Leathers for Cleaning

Taking a Western saddle apart isn't complicated. The hardest part is threading the stirrup leathers and fenders back up over the saddle tree when you reassemble it.

First, take off the stirrup bows. Then pull the stirrup leathers and fenders out; You may have to put both feet on the saddle to brace yourself. On a new saddle, the leathers are fairly stiff in the spots where they go up around the bars and bend sharply. As you pull them out, note that the stirrup leathers and fenders thread in on top of the tree bar, between the seat and the bar, and come out above the skirt.

Once the leathers and fenders are completely out, clean and condition them, then put them back over the bars. If the leather strap bumps up against a little step or shelf on the bar, you'll have to ease the leather over it; a strong, thin piece of pattern plastic used in making saddles (ask for a piece at a saddle shop) makes a great "shoehorn." Cut the plastic into a band just big enough to cover the shelf; a strip about 3 inches wide and 6 inches long, and rounded on each end so it won't catch, works well for most saddles.

With the saddle upside down, push the strip in from the top of the bar to make a smooth ramp that will keep the leather from hitting the shelf. When you thread the stirrup leather back in, it should lift up over the plastic and come out between the skirts. Then you can pull it all the way out, bringing the fender into its proper position beneath the side jockeys. Finally, remove the plastic strip and double the stirrup leather back around, pulling it down beneath the saddletree bars.

Training New Stirrups

The stirrup leathers on a new Western saddle may be stiff and out of position for your feet and legs. They may hang flat against the horse, so that your toe sticks into his belly when you try to mount, unless you turn and hold the stirrup with your hand. When you swing into the saddle and try to find the stirrup on the other side, you may have to feel around with your toe to locate it. You may not have time to fiddle with the stirrups if you're having trouble with the horse or are training a young horse. Additionally, stirrups that hang too straight put pressure on the outsides of your feet and are hard on your ankles and knees.

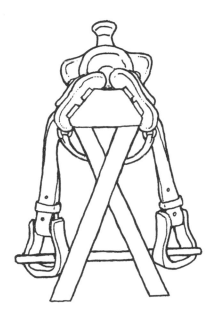

Insert a broomstick through the stirrup bows.

An easy way to solve the problem is to dampen the portion of the fenders and stirrup leathers that needs to be turned. Twist the stirrups into the proper position, give them another half turn, and insert a broomstick through both of the stirrup bows to keep them turned. If you keep the fenders twisted when you're not using the saddle, they'll bounce back to a happy medium that's more comfortable for your feet when you remove the broomstick. Once the saddle is broken in, you won't need to use the broomstick anymore.

Girths and Cinches

Some horses have tender girth skin that rubs raw easily, especially when the horse is fat and out of shape. Taking the horse for short rides is a good way to toughen up girth skin gradually, though a horse may still develop a sore during early spring rides. Using fleece or another soft material as a girth cover helps, but doesn't necessarily prevent a sore.

To reduce the risk of irritation, use a neoprene girth or cinch, which is usually made of nylon webbing covered with a sleeve of neoprene, the soft, rubberlike material used in a diver's wet suit. With this nonabrasive girth, even a raw sore will often heal while the horse is being ridden.

Another advantage to neoprene is that it can be washed quickly and easily just by running water over it. Because it doesn't absorb moisture, it dries within minutes (or you can dry it immediately with a cloth). You can have it completely clean each time you use it. If you use the same girth or cinch on more than one horse, wipe it with a mixture of water and Nolvasan (chlorhexadine) to prevent the spread of fungal infections, such as ringworm or girth itch.

Staying Undercover

Place a dust cover over your saddle when it's not in use. Besides keeping your tack cleaner, it prevents the leather from drying out. Bright sunshine (even through a tack room window) will dry and fade leather over time. Don't use a waterproof covering like plastic, which holds in moisture and promotes the growth of mold. Instead, choose a fabric that lets the leather breathe. An old, fitted twin bedsheet works nicely; the saddle skirts will hold the corners in place.

When storing a saddle, don't overwrap it or use an airtight covering, or moisture may condense and collect inside. Prolonged exposure to moisture will ruin the rawhide tree of a Western saddle.

Storing a Saddle on a Rack

A good saddle stand or rack is best for storing a saddle, because it keeps the saddle leathers in proper position. Cowboys often hang a saddle by its horn from a rope in the barn (to keep it off the ground and away from mice). Although fine between daily rides, it's not a good idea for long-term storage: The horn is up and the saddle is twisted, the stirrup leathers hang too far back and may be permanently bent or creased, and the fenders and side jockeys stick out like wings and may become set in the wrong position.

Use a stand or rack that lets the saddle rest on its bars and not its center, just as it would on a horse. The top of some racks consist of a small pole or rod, or two boards put together like a peak roof. If the stand is too narrow or the top too high, the weight of the saddle may fall on the rod or sharp peak rather than the bars. The stand or rack should be flat enough so that the top doesn't fit up between the bars.

Never put a saddle on or under a wet saddle blanket; hang the pad or blanket separately to dry. (For instructions on building your own saddle racks, see pages 49 to 51.)

Saddle Hanging Tip

If you do want to store a saddle in the short term by suspending it from a rope, make sure the saddle hangs from the fork rather than the horn. This puts the rope closer to the center of the saddle's weight, balancing it in a nearly level position (instead of tipping it upward) and allowing the stirrup leathers and skirts to hang straighter. To do this properly, make a nonslip loop in the end of the rope (a bowline knot works well; see page 240) and push it through the fork from back to front and then up over the horn.

Portable Saddle Rack

Here's a simple, sturdy saddle rack that you can easily take apart for storage or for transport in a car trunk or horse trailer. All you need are a 2-foot-long 2 X 4 for the top, a 2-foot-long 2 X 12 for the bottom, and a 26-inch-long pipe with a 1-inch diameter and threaded at both ends to connect the top and bottom. To secure the pipe to the boards, screw a 1-inch pipe flange to the center of each board. You can then  quickly screw the pipe into the flanges to assemble the saddle rack, or unscrew it to take it apart. To make the rack fit a saddle even better, glue or screw a tapered piece of 2 X 4 to one end of the top board (for the front of your saddle), then pad the top board with cushioning material.

Storing a Saddle on the Floor

If you keep your saddle on the floor, set it on its side on a clean pad or blanket so that the weight is distributed evenly and the skirts and stirrup leathers are straight. Resting a saddle on its front with the seat up causes the fenders to splay or wad up underneath, and the side jockeys to stay in an abnormal position after storage.

Using Is Better Than Storing

Lack of use can be hard on a saddle, unless you take pains to protect it during storage. Tack used regularly stays more pliable and receives more attention, because you can see problems and address them right away. Store your saddle properly and in a place where you can see it — remember, a saddle out of sight is out of mind. Don't leave it in an outlying shed where birds and mice have access to it.

ADVICE *for*
LONG-TERM SADDLE STORAGE

Take a little extra time to store your saddle properly.

★ Always clean a saddle thoroughly before long-term storage to remove dirt, fir needles, or other foreign materials from under the skirts. Pull off the skirt, lift up the front seat jockeys, and take the back jockeys off and clean underneath them. It takes two or three hours to do a good cleaning job if you do it yourself. Or you may want to consider taking the saddle to your local saddle shop for a full cleaning and oiling, and for replacement of any worn parts, such as the latigo, sheepskin lining, saddle strings, or treads. If your saddle gets hard use, taking it to a shop is a good idea now and then, even if you're not storing it.

★ Don't forget to thoroughly clean the stirrup bow treads. If there's any manure on them, the acid will eat the leather.

★ If it's a Western saddle, put a short broomstick through the stirrup bows to keep them in position (see Training New Stirrups on page 97). Some saddle stands have a hook that holds the stirrup leathers in place.

★ Take off the cinch, then wash it and hang it straight. Also remove the rear cinch, and store it flat or hanging, so that the leather won't permanently curl.

Storing Bridles

A bridle will keep clean and dust free if you store it in an old pillowcase. Arrange the headstall and coiled reins inside the case, then hang the covered bridle by its crownpiece over a rounded hanger or a small can nailed to the tack room wall. Don't hang a bridle over a nail or narrow hook; the headstall will lose its shape, and the leather will bend and possibly crack. To make your own bridle hangers, see page 52.

Repelling Rodents and Moths

Rats and mice will chew on leather and quickly ruin it. And if any of your tack is lined with sheepskin, you may need to protect it from moths.

★ Mothballs are good insurance against moth damage, and they repel rodents, as well. Fill small plastic bags with mothballs (about a half dozen in each) and locate them next to your saddles and between your saddle pads.

★ Some saddle-care products claim to repel rodents and insects. When storing your saddle, use that type of product instead of neat's-foot oil, which is edible and attracts rodents. Don't leave anything tied to your saddle.

★ An old cedar chest makes a great tack trunk and can be relatively inexpensive if you find one at a yard sale or auction. Not only will it keep mice out, but the cedar will also deter moths and other insects.

★ For more tips on managing rodents in your barn, see pages 54 to 56.

Fitting the Bit

A horse's bit should fit him properly, or it'll annoy and pain him, making him throw his head in the air or tuck his chin to lessen the pressure. Be sure the bit is the right width: Too wide and it'll slide back and forth; too narrow and it may pinch the mouth and cause sores. Measure the width

TACK

of your horse's mouth by pulling a string through it where the bit goes, then mark the string or tie a knot in it at each side of the mouth. Check the string against the bit you're using or thinking of buying. The mouthpiece should be about $1/2$ inch longer than the string, so that the hinges of the snaffle rings will just clear the corners of the mouth and a curb or Pelham won't rub the corners. A bit that is slightly (but not too much) wider than the mouth won't rub, pinch, or slide around.

Adjusting the Headstall

Proper bit fit depends not just on the width of the mouthpiece but also on how the headstall is adjusted. If the cheekpieces are too tight, the headstall is likely too small for the horse, and the bit may be pressing against the cheek teeth (the grinding teeth at the back of the mouth) and rubbing the corners of his mouth, causing sore spots. If the headstall is too large, the bit will hang too low in the mouth and may clank against the front teeth. If the horse gets his tongue over the bit, he'll avoid the pressure and ignore your signal to slow down or stop.

With a curb bit, the mouthpiece should rest comfortably at the corners of the horse's mouth, with no wrinkle in the skin. With a snaffle, there should be just one small wrinkle. When you start a young horse in a snaffle (his first bit), make it a little tighter (one and a half or two wrinkles) for the initial few sessions, to be sure he can't get his tongue over the bit. At first he'll try to spit out the bit and work it with his tongue. After he's accustomed to having it in his mouth, readjust the headstall so it's not so tight.

Checking the Balance on a Curb Bit

A bit with shanks should balance properly in the horse's mouth so it will be comfortable, exerting no extra pressure on any one area and effectively transmitting your signals through the reins. One way to test the balance of a bit you're thinking of buying is to rest the middle of the mouthpiece on one finger. If the shanks hang down where they should, then it's a well-balanced bit. An unbalanced bit won't stay in place. The lower ends of the shanks will move forward when the horse mouths the bit, altering the points of contact and interfering with the rein signals.

TECHNIQUES *for*
CLEANING METAL PIECES

Try these methods for keeping bits, stirrups, spurs, and other metal pieces clean and tidy.

★ Wash the metal part in soapy water (use any good hand or dish soap). Then polish it if you want it to really shine. A liquid metal polish works well, but polish-impregnated fiber wadding isn't as messy and won't spill on the leather portions of the tack.

★ For a fast shine on silver pieces and to eliminate tarnish, scrub with toothpaste and a little water (a toothbrush makes a great scrubber) and rinse.

★ To scour off green and brown stains on a metal bit, wet the mouthpiece and rub sand on the crusty areas with your hand or a rag. Be sure to rinse away all the grains afterward.

★ To clean and sterilize a bit without scrubbing, dip it in a saucepan of boiling water to which you've added $1/2$ cup each of baking soda and mouthwash.

★ Make a habit of cleaning the bit after each ride to get rid of dried saliva, grime, and the remains of chewed grass or hay. Dunk it in water and wipe it dry with a clean towel or cloth, though even a quick wipe with a damp cloth will do. Avoid getting leather reins or cheekpieces wet, but if they're nylon, it won't matter.

HALTERS

Halters should fit comfortably and be strong enough to never break when the horse is tied.

★   ★   ★

Fitting a Halter

A halter that's too small and tight may chafe the horse and cause a sore, or restrict his jaw movement when he chews. If too large and loose, it may slip off when he rubs his head or pulls back when he's tied. He may also catch an oversize halter on something, or snag a hoof in it when he tries to scratch an ear with a hind foot.

On a halter that fits properly, the crownpiece will rest comfortably over the top of the head behind the ears, with enough room for you to slip a finger under it. The throatlatch will be just loose enough for you to slide your hand under it at the side, where it meets the cheekpiece. The noseband should hang about 1 to 2 inches below the horse's cheekbone, with enough space to fit your fingers between it and the horse's nose.

Adding More Holes to a Nylon Halter

If you need more buckle holes in nylon tack, a leather punch alone won't work because a cut surface tends to fray and break down. Use a leather punch to make the pilot hole, but finish by melting the material around it from both sides with a soldering iron or wood-burning tool.

If you don't have either of those implements, you can heat a nail of the proper diameter. Wearing thick gloves or an oven mitt, hold the nail by

---

SOFT NOSEBAND

To make an inexpensive, homemade noseband cover that won't rub the horse's face while longeing, cut the sleeves off an old sweatshirt and fit one inside the other to create a double thickness. Stitch around the cut edges so they won't unravel. Add buttons (be sure to put them on the outside) and buttonholes for fastening the material around the noseband.

DON'T TIE WITH A LEATHER HALTER

Leather halters are nice for show but not strong enough to use for tying. When exposed to sweat and weather, leather tends to break down and weaken. A halter made of good nylon rope or nylon webbing is generally better for tying a horse.

the head with pliers while you heat the tip over a safe source of high heat. In addition to creating holes, a hot nail comes in handy for cleaning up an existing hole that is frayed and hard to buckle.

Getting Rid of Frayed Halter Edges

If a nylon web halter starts to fray along the edges, machine wash it in a pillowcase. While it's still damp from the spin cycle, singe all the frayed surfaces and edges with wooden matches or a cigarette lighter. The heat will burn off the frizzy nylon without harming the wet webbing.

Making a Breakaway Halter

A halter should never be left on a horse full-time, except in unusual situations. And in those instances, it should always fit properly and have a breakaway feature. You can turn a regular rope halter into a breakaway halter just by adding a twine loop to the fastening end, so that the hook goes in it instead of into the rope loop. Readjust the halter so that it fits properly with this extension. If the halter ever catches on anything, the horse can pull free because the twine fastener will break.

Use this halter only when you're leaving the horse loose and not when you're tying him, because he may learn that he can break free just by pulling back. When you tie a horse, always use something he can't break (see chapter 9 for information on knots and ties).

BREAKAWAY WEB HALTER

You can alter a nylon web halter to make it into a breakaway halter. Stitch a piece of thin leather (like a worn-out belt or some other piece of old, decaying leather) under the crownpiece, and punch holes in it that correspond with the holes in the nylon halter. Buckle the leather piece instead of the nylon one when you want it to be a breakaway halter. Tuck the end of the nylon piece through the side of the buckle to keep it from flapping around. The old piece of leather will be strong enough for leading the horse, but will break if the halter catches on something. When you want to tie the horse, unbuckle the leather strap and rebuckle the nylon one.

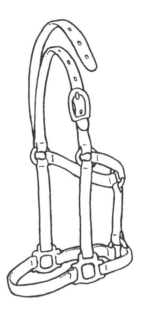

Bringing a Halter When You Ride

Whenever you ride outside an arena, have a halter and lead rope in case you need to tie your horse. The simplest and safest thing to do is leave a halter under the bridle. You can carry snap-on lead rope coiled and tied to the saddle or stored in a saddlebag. Another option is to use a halter with a lead rope attached; loop the rope around the horse's neck and tie it back to itself just behind the halter. When you dismount for any length of time you'll be able to tie, hold, or lead the horse more easily. If a horse must be led from another horse, it's always easier to do it with a halter rope than with bridle reins.

BLANKETS, SADDLE PADS & BANDAGES

Saddles, bridles, and halters are not the only equipment that requires proper fitting and care. Blankets, pads, and even bandages should be well matched to the horse as well as the occasion.

★ ★ ★

Fitting a Blanket

It's important that any blanket you use on your horse fits properly. If it's too tight, it may rub and irritate the skin and cause sore spots. A high-withered horse may end up with a sore at the top of his withers. A blanket that's too large and loose may slip and end up under the horse's belly. When putting on the blanket, set it well forward and then slide it back into place, so it won't ruffle the hair the wrong way. Make sure the straps are always properly adjusted. If a horse's legs become entangled in the straps, he may injure himself in his struggle to pull free.

Routine Blanket Care

Brush all hay, wood shavings, manure, and mud off a blanket before you take it off the horse. Use a brush with very stiff plastic bristles reserved for blankets only, never for grooming. If a blanket is wet or damp, hang it over a fence and turn it over once while it's drying, or drape it over two portable saddle racks so both sides will dry.

A horse with ringworm, girth itch, or any skin infection should have a blanket that is never used on another horse. Even after his skin problem clears up, fungal spores may be transmitted to another horse.

Blanket-Washing Tips

Launder a horse blanket regularly, to extend its life.

★ Keep blankets clean; wash them at least twice a winter. Dry cleaning won't remove odors, the heat may shrink the bindings, and solvents may damage a waterproof coating.

★ Thin nylon outdoor blankets can be machine washed and dried. Use cold water, soap, and a disinfectant, and put the blanket through several rinse cycles to make sure there's no soap residue.

★ Hang cotton and wool blankets to dry, rather than putting them in a drier, where they tend to shrink.

★ Washing weighty or bulky blankets in a heavy-duty washer at a commercial laundry is usually better than trying to do it in a home washer.

When to Blanket

A horse with a good coat of hair doesn't usually need a blanket in cold weather, but he may if he's been clipped or brought from a warmer region to a colder one. Keep in mind that blanketing a horse that's unaccustomed to it may cause him to overheat and start sweating, then chill because he's damp. Don't blanket a horse for turnout in the early cool morning if you won't be there to remove the blanket if the weather warms in the afternoon. To tell if a horse is getting too warm under his blanket, feel his chest or just behind his elbow; these are some of the first places he'll start sweating.

INSTANT BLANKET

If you don't have a good warm blanket for a sick or cold horse, you can improvise one with an old sleeping bag. Drape the sleeping bag soft side down over the horse, leaving the durable surface on the outside. Cut two slots at the front for securing the "blanket" under his neck with baling twine or any adjustable straps you have on hand. Cut two slots behind his elbows for securing a strap under his belly at the girth area. If the horse will be moving around much, add another strap across the back of the blanket, beneath his buttocks.

QUICK FIXES *for*
BLANKET REPAIR

Many blanket fixes are much easier than you may imagine.

★ A quick and easy way to fix a snag or rip is to iron on a denim patch (the kind you use on jeans). For extra strength, overlap the patches. And if you want to make sure they stay in place for a long time, stitch around the edges. If there's an electrical outlet in your barn or stable, you can use a flat bench as an ironing board.

★ When sewing a rip in a blanket, use the heavy thread a saddlemaker uses for leather tack. If you don't have any, you can use waxed dental floss, which is much stronger and wears longer than regular thread.

★ You can replace a missing or broken chest strap on a horse blanket with a nylon dog collar. Not only are they fairly inexpensive, they're usually stronger than the original blanket straps.

★ If the fastener on a horse blanket tends to come loose when the horse is turned out, hold the fastener together with a plastic twist tie (the kind used on bread bags and other grocery items).

Nonslip Saddle Pad

To secure a saddle that tends to slip either back or to one side when you mount or during a ride, place a slightly damp leather chamois (the kind used to wash and polish a car) under the part of the saddle pad sitting over the horse's withers. (You may have to trim the chamois to fit, without showing, under an English saddle.) The damp leather will "stick" to your horse's coat and to the saddle pad, holding your saddle perfectly in place for short rides, such as in the show ring. However, don't rely on this solution for an all-day trail ride.

Start with a chamois about 2 feet long and 1 foot wide. Soak it in water and then wring it out, leaving it damp but not too wet. Fold it evenly in half, forming a 1-foot square, and place it over the horse's withers with the fold toward the front. Make sure it's completely smooth (wrinkles may cause sores), then carefully place your saddle pad and saddle over it.

Another Nonslip Tip

If your saddle pad or blanket tends to move backward on your horse, working its way behind the saddle when you ride for long days in steep country, try this simple trick to keep it in place. Punch a smooth hole in the front of the pad, then loop a leather thong or nylon shoestring through the hole and over your saddle horn. The thong should work well if you loop it while the pad and saddle are positioned properly on the horse's back and if there is not too much slack.

Use a leather thong to keep a saddle pad from slipping.

DRYING RACK FOR SADDLE PADS

If you don't have a good place to hang wet saddle pads and blankets, put up an old curtain rod big enough and sufficiently sturdy to hold several of these items. Choose a handy spot in or near the barn or tack room and, for quicker drying, be sure there's enough clearance for air to circulate under and around the pads and blankets.

Washing a Fleece Saddle Pad

Fleece pads are not only comfortable for the horse, but also easy to keep clean and soft because they're washable. However, they can be a nuisance to clean in a washing machine, leaving horse hair and grime in the tub and requiring several rinses to get rid of soap (the residue can irritate a horse's back).

The type of power washer used for cleaning motors and farm equipment is handy for washing fleece saddle pads. Even without soap, the pressurized stream of water can lift out dirt and hair better than a washing machine. Drape the dirty pad over a sawhorse or saddle stand for washing; after cleaning both sides, hang it on a fence to dry. You can also use the pressure washer at a car wash.

Whichever cleaning method you used, you may want to fluff up the pad after it's dry, using a vacuum cleaner. This will pull out any hair left in the pad.

Cleaning a Fleece Pad between Washings

After you've used a fleece saddle pad, let the sweat dry, then brush the pad with a steel-bristled dog-grooming brush or a sheep-grooming card. You'll remove dried sweat and caked dirt or hair while fluffing up the fleece. A quick brushing between rides makes the pad more comfortable for the horse and lets you extend the time between washings.

Preventing Saddle Sores

A saddle that doesn't fit the horse may put too much pressure in one spot instead of evenly distributing your weight. You'll know there's a problem if you see a dry spot at the pressure point when you remove the saddle, even though the rest of the back is sweaty under the pad. The extra pressure there interferes with the skin's ability to sweat.

If you can't change to a better-fitting saddle immediately, you may be able to alleviate the pressure by altering a saddle pad. Select a thick pad, then carefully determine where the pressure spots are when the horse is saddled (the sides of the withers are common locations). Cut away the portion of the pad that corresponds with each spot. This can prevent sores, hair growing in white, and thick scar tissue.

You may have to experiment to see if one holey pad resolves the problem or if you need to use two. A double-thick pad will raise the saddle and distribute its weight over a larger area. In either case, make sure the holes line up with the pressure areas on the horse's back.

Homemade Leg Bandages

You can make protective, washable leg bandages with 6 yards of quilt batting and four old standard-size pillowcases. For each pillowcase, cut a piece of batting the same size as the case and fit it inside, then machine-sew it across the open end. Next, sew a swirling pattern completely through each leg wrap to secure the batting to the pillowcase in several places. This will prevent it from wadding up or shifting position. These homemade leg wraps roll smoothly around a horse's leg and are easy to secure with stable bandages.

SAFETY TIP FOR BANDAGE PINS

If you have to resort to safety pins to hold a leg bandage in place, cover each pin with duct tape so the horse can never accidentally open it. Fold a small corner of the tape so it sticks to itself, creating a tiny tab that allows you to pull off the tape when it's time to undo the pin.

REPAIRS & GENERAL MAINTENANCE

Fixing your own equipment can save you a lot of money and, in most cases, it's easier to do than you might expect.

★　　★　　★

Riveting Broken Tack

Whether you fix broken tack yourself or take it to a leather worker or saddlemaker for repair may depend on how expensive or fancy the tack is, or if it has sentimental value. In many instances, a do-it-yourself fix is just fine.

For example, a cheap leather headstall may just have rivets or Chicago screw fasteners at the ends of the cheekpieces to hold the bit in place. If any come loose, an inexpensive pack of rivets and a simple riveter are all you'll need for a quick repair. You can also fasten broken pieces of leather together with rivets and a piece of scrap leather.

Riveting tool

One type of rivet goes in with a leather punch and a hammer. The rivet is a small, tapered copper post with a flat, round lid on the larger end and a flat washer that fits onto the smaller end. It comes in various lengths, so you'll want to choose the proper one for the thickness of the leather pieces you plan to rivet together. Here's how to put in rivets.

1. In the leather pieces punch a hole just large enough to force the rivet through.
2. Line up the holes, push the rivet into them, and place the washer on the end that comes up through the leather. About $1/16$ inch of the rivet should stick up beyond the washer.
3. With the leather on a smooth, firm surface, pound that little piece of rivet down against the washer, all the way around, flattening it out to hold the washer securely in place.

With the second type of rivet, you don't need a washer. Just position the hollow rivet in a special riveting tool (shown opposite) and press down on the handle of the riveter to force the rivet through the leather. Doing so squashes the rivet and forms a neat, splayed brad on the other side.

Stitching Repairs

For a smoother, nicer-looking repair, you'll want to sew some items of tack rather than rivet them. For example, if a headstall breaks where it goes through, a riveted area may be too thick to go through the buckle. Also, because the leather already broke once, you don't want to weaken it further by punching a hole in it for a rivet. The repair will be less bulky if you thin the ends of the broken piece and the scrap leather piece you plan to add so they'll be the proper thickness when you stitch them together.

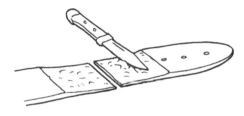

Thin the ends of the broken leather.

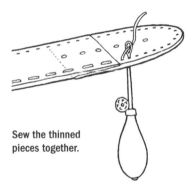

Sew the thinned pieces together.

Tools for Sewing Leather

With an awl, needles, edger, and thread, you can do nearly as good a repair job as a saddlemaker, and at lower cost. Most leather workers charge a minimum shop fee for even simple repairs. For the same price, you can buy the materials needed to do an uncomplicated repair yourself, and have them for future repairs. Because the equipment is small enough to fit in a box in your tack trunk or saddlebag, you can take it along when you travel with your horse. In your kit, don't forget to include beeswax to rub over stitches to keep them from rotting or wearing out.

Awl. It's almost impossible to push a needle through leather without first making a hole for it to pass through. An awl is a simple sharp-tipped tool that punches a small hole in leather. You can use a regular awl or a speedy type.

★ Most saddlemakers use a regular awl, which has a wooden handle. With this type of awl, you use two harness needles, one on each end of the thread. As you sew you tie each stitch with a knot, so the stitching holds better than when done by a speedy awl.

★ A speedy awl simplifies the sewing job by incorporating a hollow area in the handle for a bobbin and thread. The thread comes up through a hole in the base and passes through the eye in the tip of the large needle. When you poke a hole in the leather with this tool, the thread goes through with it, creating a lockstitch similar to one made by a sewing machine. You can order a speedy awl through your local saddle shop or a mail-order catalog for leather workers.

Edger. Before you start stitching, use this tool to make a groove for the stitches so they'll be slightly below the surface of the leather and less apt to rub or catch on something and break. The groove will also ensure that the stitches follow a straight line. (If you don't have an edger, dampen the leather before you sew it and pull the thread very tight.)

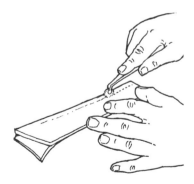

Stitch marker. A model that marks five or six stitches per inch is adequate for most repairs. That's enough stitches for a good hold, without them being so close together that they cut through the leather.

Thread. When using a regular awl, your thread should be at least one and a half times longer than the length you'll be stitching. Use even more thread if the leather is thick. Choose waxed linen thread when using a regular awl; the lubrication makes it easy to pull the thread through leather. You can thin the end of the thread with a knife and then wax it into a point for easy threading through a harness needle. Nylon thread works best in a speedy awl, because it's less binding and pulls through the instrument more readily.

Sewing with a Regular Awl
Follow these steps to create a series of strong, neat stitches.

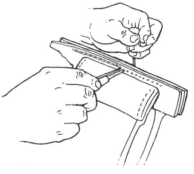

1. Thread a needle on each end of a length of thread, align the two pieces of leather, then punch a hole through the two layers with the awl.

2. Insert one needle through the hole and even out the thread on both sides. Punch the next hole and push the same needle through the back of the second hole.

Punching holes in the leather

3. Now insert the other needle through the second hole from the front side. Loop the back thread over the top of the needle as it comes through, making a twist on the back side and locking the stitch with a single knot.

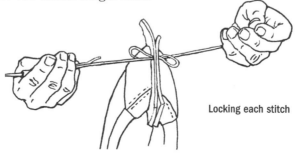

Locking each stitch

4. Continue to the next hole, pushing each needle through from the side it is on.

5. When you come to the end, do a backstitch (just put the needles back through the last hole a second time) to make sure the sewing will hold. Snip the ends of the thread flush with the leather. To make the stitching look professional, rub a little beeswax on it and push the wax into the hole to seal it.

Sewing with a Speedy Awl

With a speedy awl, you push the needle through the hole, pull it back slightly to make a loop of thread, and then put the back-side thread through the loop. Next, pull the needle and the back-side thread in opposite directions so the stitch locks together between the pieces of leather. If it is not well buried, it is more apt to catch on something and break.

Push the needle through the hole.

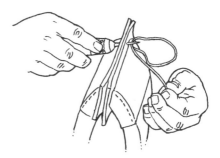
Pull back slightly to make a loop.

DON'T USE A KNIFE

When a new piece of leather tack needs another hole to fit properly, the worst thing you can do is carve it with a pocket knife. Even if you're careful, it won't make a smooth, clean hole, and it will compromise the strength of the leather in a critical place. There's also the risk of slipping with the knife and making a longer or bigger hole than necessary. Always use a leather punch.

Make-Do Tack Repair

If the stitching on a bridle or halter comes loose and you don't have an awl, you can repair it as long as you have these items on hand: a straight upholstery needle (large enough for thick leather or nylon webbing) and heavy waxed thread, fishing line, or waxed dental floss. Work slowly and handle the needle gently to keep it from breaking. If you're having a hard time pushing the needle through the leather, move it forward with pliers. Here's what to do.

1. Wrap a little electrical or duct tape around the needle where you'll hold it, to protect your fingers and provide a comfortable grip.

2. Remove the loose or broken stitches from the tack, and tie knots in the ends of the remaining stitches so they won't unravel.

3. Cut a length of thread long enough for the job, or leave the thread attached to the spool to avoid a splice. The end that's pulled through the needle should be about a third of the total length you think you'll use.

4. Starting from the outer surface of the item you're repairing, push the needle, eye first, down through the first unstitched hole until you can grab hold of the short end of the thread from the underside. Pull it through and back the needle out of the hole.

5. Still working from the outer surface of the leather, insert the needle again, eye first, partway into the next hole. Pull it back toward you a little so that the doubled thread on the underside loops open. With your finger, pass the shorter loose end of the thread through this loop, then back the needle out of the hole. Repeat the procedure at each hole you're restitching.

6. At the last hole, push the needle partway through and cut the thread, leaving enough to tie the two ends in a neat, flat knot. Snip off the extra thread at the tie.

Quick Fix for a Broken Rein

If you make the mistake of tying a horse with his bridle, or if you allow him to graze with his bridle on and he steps on a rein and then jerks his head up, there's a good chance you'll end up with a broken rein. Often, the break is very close to the bit. Here's an easy way to repair it.

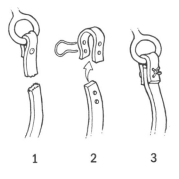

1. Undo the screw or thong that holds the rein together at the bit.

2. Punch matching holes in the end of the piece that broke off, then sandwich it between the first piece, lining up all the holes. If there is only one hole in the first piece, punch an extra one for a more secure joint.

3. Tightly screw or lace the two pieces together with a bit in place.

1   2   3

Curb Strap/Chain Repair

Occasionally, a leather curb strap breaks or a curb chain comes loose and falls off. If you're on a ride when this happens, you'll have no control when asking the horse to slow or halt. But with baling twine or a light rope you can improvise a curb strap.

Attach each piece to one of the curb rings by doubling it, putting the loop end through one of the rings, then running the ends back through

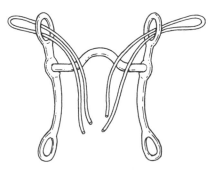

Put the loop ends of the twine through the curb rings.

Tie the attached twine pieces together.

the loop. Make a double half hitch (see page 244) to knot the strands together 2 inches or so from the bit ring. Add the second piece to the other ring in the same way. Finally, use a square knot (see page 238) to tie the ends of the two pieces together under the jaw.

A saddle string is usually too bulky for this repair but can still work if you use just one long piece. Thread it through the bit rings and tie it together with a square knot under the jaw.

TOWEL TACK BAG

A laundering sack made from an old bath towel will cushion your tack hardware better than a pillowcase. Just fold the towel in half and sew two sides closed, then add a drawstring or Velcro to the open side.

Revitalizing Velcro

Velcro closures on horse gear, such as splint boots and other protective leg wraps, sometimes reach the point where they no longer stick well. A simple way to clean up old, tired Velcro is to run a pet's flea comb through the hook (or rough) side of the closure; it removes dirt and fuzz, and often enables the closure to work properly again.

Washing Nylon Tack

Nylon tack, such as halters, headstalls, reins, and lead straps, can be put in an old pillowcase that you tie shut with string and then laundered in a washing machine. The pillowcase should trap most of the dirt and any hair, and also protect the washer from the halter hardware. It also keeps the tack from getting into a big tangle.

Put the filled pillowcase in the washing machine with a load of other laundry (such as jeans or other clothing that is not fragile or white). When the load is finished, remove the tack from the pillowcase and give it an additional rinse under a faucet or hydrant to get rid of any soap residue, then hang it outside to dry. Turn the dirty pillowcase inside out and rinse it well before adding it to the regular wash.

Fixing Frayed Nylon Rope

If a nylon lead rope becomes frayed on the end, use a burning match to melt the frizzy threads back into the rope. Work over a sink, in case you have to drop the match or douse a hot spot with water.

Avoiding Inferior Snaps

A rope or lead shank is only as strong as its snap, and a cheap snap is dangerous. Iron snaps are usually much stronger than ones made of zinc or an alloy. Some hardware is made of stainless steel, nickel, or brass, and any of these is stronger than die-cast hardware. The die-cast process traps air in the metal when the molten metal is poured into the mold, making the finished product somewhat porous and brittle. In cold weather it may become even more brittle and break under stress.

There are many kinds of snaps and hardware used in horse tack, and it isn't always easy to tell the metals apart. When in doubt about the quality or strength of an item, don't buy it. If you want to make sure the hardware is iron instead of die-cast zinc, check it with a magnet. A magnet is attracted to iron or steel, but not to zinc.

Fixing a Broken Snap

If a snap on a lead rope breaks, cut it off. Thread the cut end of the rope through the base ring of a new snap and fold it back on itself. Slide a link of heavy chain over the rope until it rests at the base of the snap and holds the doubled rope together. Tie a knot in the short end of the rope right behind the chain link, then hammer the link flatter so it becomes very tight. The link, along with the knot just behind it, will keep the rope from pulling out of the snap.

CHAPTER FIVE

Grooming

Regular grooming not only promotes a healthy coat
and hooves, but it also helps accustom your horse to daily
handling and teaches him patience and good manners.

GROOMING BASICS

To adequately groom a horse you need only a few simple tools, a bit of elbow grease, and a little know-how.

★   ★   ★

Currying & Brushing

Typically, it is most efficient to begin a good grooming at the neck, working back to the hindquarters. On areas other than the face and legs, use a currycomb or a stiff brush to move dirt and dandruff to the surface. A rubber currycomb applied in a circular motion will massage the body, increase circulation, stimulate oil glands, and bring up some of the deeper dirt and dandruff next to the skin. The firm rubber knobs work well over muscle, but can cause discomfort over bony areas. On a horse that has really sensitive skin, use a grooming mitt (a rubber one with small nubs). When brushing, go with the lay of the hair, using firm strokes.

Next, you can use a soft body brush on the entire horse to lift out surface dirt and dust. If the horse is sensitive about having his face brushed, use a damp towel or soft cloth to clean around his eyes and nose (and any other sensitive places, such as the sheath or udder).

GROOMING TOTE STAPLES

Keep the items listed here on hand for routine grooming sessions.

- ★ Metal currycomb (for removing matted winter hair)
- ★ Rubber currycomb (for regular grooming)
- ★ Stiff-bristle brush
- ★ Soft-bristle body brush
- ★ Rubber mitt
- ★ Hoof pick
- ★ Towel or some other soft cloth (for cleaning sensitive areas)
- ★ Knit gloves with rubber reinforcement on the palms and fingers (to keep your hands clean while grooming)

Grooming-Tool Carriers

Make grooming quicker and easier by wearing a work belt or apron with pockets for holding a hoof pick, brushes, and other equipment you're using. Another option is to organize your tools in the compartments of a plastic tote or caddy.

Handy Grooming Cloths

Old socks make great grooming or tack-cleaning cloths that can be washed and reused many times. Pull the socks over each other to form a ball that you can store in your tack room, car trunk, or horse trailer. When you need a soft cloth to rub off dirt or apply a coat-shine product in places too sensitive for a brush, just pull a sock off the ball and slip it over your hand as a mitt. For larger, flatter cloths, cut each sock down the center (or at the holes); store them in a flat pile in a plastic zippper bag or other container. A car-washing mitt makes another handy tool for wiping dust off the coat, and it's sometimes easier to spot in your tack box than a grooming cloth.

Cleaning & Drying with Sawdust

A good way to clean a horse on a cold day or when he's sweaty after a ride is to rub several handfuls of clean, dry sawdust into his coat, then brush it out. The sawdust absorbs moisture and removes dirt. If your horse is very sweaty, first rub his coat with a towel so the sawdust dries faster and he's less likely to catch a chill.

KEEPING BRUSHES CLEAN

Remove dirt and hair from a brush during grooming by rubbing it firmly against a tight wire or net wire fence, or by rubbing it with a curry-comb. To prevent a brush from filling with hair so quickly in the future, apply a commercial hair coat conditioner on the clean bristles after washing it. Regular dish soap is fine for washing brushes. After rinsing them thoroughly, rest them on their sides in the sun so the wood handles will dry faster.

To avoid spreading skin problems from horse to horse, allot each horse his own grooming tools, especially a brush and a curry comb. If you suspect a problem, use a separate pair of gloves for that particular horse or wash your hands between groomings.

USING A HAIR DRIER ON A SWEATY HORSE

Some riders clip a horse for winter riding, to save time on drying a sweaty mount. However, a clipped horse must be blanketed to protect him from cold weather. An alternative is to use a hair drier for after-ride grooming and drying. Horses quickly get used to the noise and enjoy the warm air, which will also feel very pleasant on your cold fingers after you ride.

Be sure to use nontoxic sawdust. Some woods can have adverse effects. Black walnut, for example, can cause laminitis, and yellow poplar can make your horse itchy.

Removing Manure & Grass Stains

A simple way to clean a stain from your horse's hair is to mist the area with rubbing alcohol you've poured into a clean spray bottle, and then to rub the stain with a cloth. You can also pour a little vinegar directly onto a cloth and use it to rub off the stain. When you're done, rinse the area with water.

Yet another way to get rid of stains on white markings is with a little bran or soybean meal mixed with just enough water to make a paste. Wet the stained area, then smear on the paste and scrub it into the hair with a small brush, such as a fingernail brush or toothbrush. Leave the paste on for several minutes, allowing it to absorb the stain. Then hose or sponge it off and dry the hair with a towel.

Brightening White Socks

Make your horse's white socks or stockings even whiter by rubbing a little baking soda, corn starch, or baby powder into the hair and then brushing off the excess.

REMOVING BOT EGGS

There are several ways to eliminate bot eggs (laid by the botfly and stuck on the ends of a horse's leg and flank hairs.) You can gently rub them with fine sandpaper, going with the lay of the hair. Another option is to pull the eggs off with the serrated edge of a curved grapefruit knife. A third method used by some horsemen is to gently run a disposable razor along the direction of the hair to cut off the eggs; be especially careful around the bony regions, such as the knees and hocks,

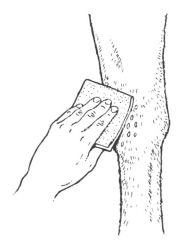

Removing bot eggs with sandpaper

to avoid nicking the skin. Yet another is to hold a very warm, wet cloth against them for several minutes, which will stimulate the eggs to hatch and the larvae to attach themselves to the cloth.

Creating a Final Sheen

To give a final cleaning and shine to your horse's coat, spray a little water on your grooming cloth so that it's barely damp. That little bit of moisture will remove the last of the dirt and dust sticking to the hairs. Alternatively, you can use a solution of 1 part liniment to 10 parts water; besides collecting dust on the cloth, the liniment will quickly evaporate from your horse's coat.

Getting a Horse Used to Clippers

Before clipping a horse, hold a few practice sessions during which you just turn on the clippers without doing any cutting. Once you know he's not afraid, hold the clippers against his neck or shoulders (with your hand between the horse and the clippers), so he can feel the vibration and become at ease with it before you actually clip him. An effective way to quickly train a nervous horse to accept clipping is to use a Stableizer (see page 265). The tranquilizing effect induced by this aid will make him associate the clipping with good feelings instead of scary ones.

TIPS *for*
VACUUMING THE COAT

To suck dirt, dust, and dandruff from a horse's coat without removing the natural oils, some people use a vacuum cleaner designed for horses. This grooming method has the added benefits of keeping you cleaner. Plus it is a better choice than bathing on a cold day. You can also use the vacuum to remove horse hair from your tack and clothes. A small, cordless handheld vacuum is an inexpensive alternative. A vacuum is a real time-saver, but it's important to use it carefully.

★ Never use a vacuum on a wet horse, or you may shock yourself and the horse.

★ While the vacuum sucks, the exhaust end discharges air, so make sure you're not blowing dust from the barn floor into the air.

★ Keep the hose, cord, and machine itself out of the way so that you and your horse won't trip over or step on them.

Clipping a Bridle Path

Trimming the mane behind the ears allows the headstall or halter to lie neatly at the top of the head and neck without a wad of mane beneath it. A short bridle path of 2 to 3 inches is adequate, but when showing a horse some people trim a longer path to make the neck look nicer. The actual length will depend on what is specified in the rules for that breed.

To clip a bridle path, use a blade setting that gives the closest shave. Loosen the halter enough that you can slide it back a few inches to hold the rest of the mane out of the way. Gently fold the left ear forward out of the way as you clip along the top of the mane from back to front, then front to back, and finally down each side for a smooth, uniform finish.

Clipping Ears

It's fine to clip the long hairs around the ears for showing purposes, but it's inhumane to trim the hair inside the ears, which is there to keep out flies and debris. To trim the long hairs, fold the ear together and clip only the ones that protrude from the edges.

Leaving Whiskers Alone

A horse uses the whiskers on his chin and muzzle as feelers to know how close he is to the ground, stall wall, or bottom of a feed box. The long hairs around his eyes also serve as feelers. It's important never to trim any of them.

CLIPPER BLADE CARE

Make sure the blades on your clippers are sharp, as dull ones will pull on the hairs. Lubricate the blades (following the manufacturer's directions) to reduce heat and friction and to keep them from wearing out too fast. One drop of oil before clipping and again before storing the clippers is usually adequate.

Shedding Winter Hair in Spring

Horses start to shed their winter coats when days get longer in the spring. A horse kept in a barn with the lights on at all times won't grow as much winter hair and will shed sooner. Blanketing a horse will also stimulate quicker shedding by making him warmer. When he sweats, his body thinks it's time to get rid of all that extra hair.

If your horse's belly hair is slow to shed, cover that part of the body by attaching a burlap bag to his blanket; sew strips of Velcro to the bag as well as the underside of the blanket. The extra warmth will make the horse sweat and shed, and the burlap's rough texture will also help rub off the loose hair.

Using a Pumice Stone for Shedding

The pumice stones with which people remove rough skin from their elbows and feet are also handy for shedding hair from a horse's legs, face, and underline. Using a pumice stone during the horse's first spring bath will also help remove dirt embedded deep in his long winter hair.

Making a Scrubber from Baling Twine

If a horse has wet or dried mud in his long, matted hair when you're trying to shed him out, try rubbing it out of his coat with a balled-up piece of baling twine. Once the worst of the caked dirt is gone, you can use a brush to finish the job.

FEEDING FOR A HEALTHY COAT

A healthy horse on a balanced diet sheds more quickly than an undernourished horse that's not getting enough protein and vitamin A. Hair is made up of protein, so a horse needs an adequate amount to grow a new summer coat. Adding a couple of ounces of vegetable oil to his daily grain ration also seems to promote a healthier, shinier coat.

Along with a balanced diet, the ingredients for a good coat are regular grooming and body massage, both of which increase circulation, promote healthy skin, and stimulate oil glands.

GROOMING BASICS

MANES & TAILS

A scraggly mane and tail can greatly detract from your horse's appearance, even when the rest of his body is well groomed. Attending to these areas on a regular basis will keep them neat and free of knots.

★　　★　　★

Brushing the Mane and Tail

A human hairbrush works well on both the mane and tail. For a nice-looking mane, brush one side completely, then flip the mane over and brush it on the other side before flipping it back. A plastic hair pick for human hair also does a good job of combing out a mane, and it's small enough to carry in your pocket. Use a metal mane comb or a hoof pick to remove any snarls you may encounter.

For the tail, start at the bottom and work up gradually, brushing small sections at a time to untangle any knots. Pull a long tail over your leg so you can work on the end more easily. Unless you need to thin a tail, don't use a comb, because it tends to pull out too many hairs.

Tackling Big Knots

A horse that's been out on pasture all winter without being groomed may have some extra-big or tight knots in his tail as well as his mane. Here's how to avoid having to cut out the knots in the tail: Pour 2 cups of ShowSheen or a fabric conditioner in a small bucket, such as an ice

---

PREVENTING TANGLES

If a horse's mane or tail tends to twist, tangle, and knot (often called witch's knots), use hair conditioner on them (but not the body) when you bathe him. Rinse thoroughly to avoid creating dandruff. As an alternative, apply a fine mist of silicone spray on matted areas. It'll make the hairs slide apart more easily, reducing hair loss from routine brushing and combing.

---

cream tub, then push the knotted part of the tail into it and saturate the knot. This type of silicone-based product makes the hair slippery so that the individual hairs slide apart more readily. Don't try to brush the tail for two days, until it completely dries. Use a wire dog-grooming brush (the kind used for preparing dogs for shows), working gradually from the bottom up. After you've brushed out the tail, wash and rinse it, then apply conditioner, which will make it less likely to knot up again.

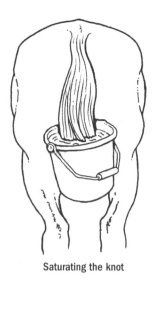

Saturating the knot

Training the Mane

Some horses have a thick, bushy mane that won't lie neatly on one side of the neck. Braiding the mane on the desired side usually solves the problem. Fairly large, loose braids will do the job; if they're too tight, some hairs may break off. Leave the braids in for a day or two, or even longer if the horse isn't rubbing them. After unbraiding the mane, brush it thoroughly, wetting the hair as you go. The mane should now stay in place.

Large, loose braids

If the mane starts to separate again, rebraid it. Usually two or three braidings are enough to train the hair, but a really thick, stubborn mane may require additional braidings over several weeks to keep the hair from wandering onto the wrong side of the neck. An especially thick mane may even need thinning to make it cooperate.

Combating the Frizzies

If your horse's mane looks frizzy after it's braided, slick down the small hairs that are sticking up with a little petroleum jelly. You can also use hair spray (which won't collect dust) or hair gel to keep the wispy hairs under control as you braid.

Thinning the Mane by Pulling

Most old-time horsemen thinned a thick mane by "pulling" it, taking a few hairs at a time and pulling them out. Some horses don't mind this practice, but others are touchy. If a horse objects, try pulling a little at a time after each ride; the hairs will come out more easily after the horse has exercised and sweated.

If you really need to thin the mane all at once, warm each section with a hot, damp towel before you pull it. If the horse still objects, spray some commercial sore throat spray (made for humans) onto the base of the hairs, to dull the sensitivity of the nerve endings. If you don't have sore throat spray, pour a small amount of horse liniment along the base of the hairs in a 4- to 5-inch-long section of mane. After you pull that section, apply liniment to the next section, and so on. Liniment seems to work best for dulling pain if applied just beforehand.

Pouring liniment along
the base of the mane

Thinning a Mane Without Pulling It

You can use a thinning comb (made for people or dogs) or a clipper blade to shorten and thin a mane. Tease the hairs, but instead of pulling them out, trim them very close to the neck on the underside of the mane. Doing so will help the rest of the mane above those hairs fall more naturally on the correct side.

Another method is to run the blade of a stripping knife (used for trimming a dog's coat and available in pet supply stores and catalogs) through the mane. To shorten the mane, stroke the knife more sharply through the ends of the hairs. The knife will produce an even trim, not the chopped, fresh-cut look that scissors would.

Removing Pitch

After rubbing his mane on a pine or fir tree, a horse may end up with globs of pitch sticking the hairs together. You can dissolve the pitch by applying fingernail polish remover directly on it with gauze or a terry cloth towel, being careful not to get any on the horse's skin; wash the mane afterward. For a less caustic alternative, try neat's-foot oil. In addition to removing pitch, it'll get out snarls if you sprinkle it on the knots before brushing them.

Getting Burs Out

To remove burs (from such plants as burdock and cocklebur) from a mane or tail,

USING PULLED HAIR AS BRAIDING THREAD

If you end up with some long hairs after thinning your horse's mane, save them. Later, when you put in braids for showing, you can use those strands instead of thread when sewing in the braids to keep them in place. Not only are the strands of hair just as strong as thread, but their color will blend in perfectly.

MANES & TAILS

135

soak them with vegetable or mineral oil to keep microscopic flakes from floating in the air and possibly into the horse's eye, where they can cause a serious problem. Wear rubber gloves to protect your fingers. After removing burs, shampoo and thoroughly rinse the hair to get the oil out.

Doing a Neat Roach Job

If your horse has a roached (trimmed-off) mane, the appearance of the horse's neck can be enhanced by careful trimming. Clip it closer at the top of the neck, using finer blades on your clippers for the bridle path (nearer the ears). For the rest of the mane, use a coarser blade, leaving the hairs just a little longer over the arch to give it a fuller look.

Bleaching out Tail Stains

If your horse has a white tail that's often discolored with yellow stains, keeping it clean can be frustrating. Shampoo or bleach usually works well, but more drastic action may be required for a tough stain. For that, you can use hair coloring from a drugstore.

First make sure the tail is clean (shampooed, rinsed, and wet but not dripping). Wearing plastic gloves and using warm water, apply the color rinse according to the directions on the package. Take care not to get any on the horse's skin, and use it only on the lower two-thirds of the tail so it won't come in contact with the tailbone. To protect the horse's legs from the chemical, put the treated tail into a large plastic trash bag. Double up the bag, tying it snugly around the tail and to the tail hairs above it so it won't slip. Leave the bag on for 20 to 30 minutes.

After removing the bag, shampoo and rinse the tail in warm water. Apply conditioner to treat the brittle effects of coloring the hair; rewrap the tail (with the conditioner still in it) for about 30 minutes. Rinse the tail, which should be beautifully white again.

Making a Tail Look Fuller

You can add some fullness to a thin tail by washing it, applying a good hair gel, and then braiding small segments along the edge. When the tail dries, unbraid those sections and carefully comb them out with your fingers, leaving the hair a bit wavy.

HOMEMADE TAIL BAG

For a quick and inexpensive tail bag to keep your horse's tail clean, cut two holes in the top of a large tube sock. Braid the tail and put it in the sock, then thread a large shoelace through the two holes of the sock as well as the braid, and tie the ends together below the tailbone. You can spray the sock with insect repellent to keep flies away.

Another option is to make a tail bag from a sweatshirt sleeve. Cut off the sleeve and make three equally spaced slits in it extending from the cut edge to about 2 inches from the cuff. Sew the strips into 3 tubes.

Sweatshirt sleeve tail bag

Put the tail inside the sleeve with the cuff at the top and separate the tail into thirds. Pull each third into one of the tubes (rolling the tubes first, as you would a long sock before putting it on, makes this easier). Braid the filled tubes. Then secure the bottom of the braid with a rubber band. This method of encasing the tail provides a lot of protection.

FEET

When grooming a horse, cleaning the feet is an important part of the job. Besides picking out packed dirt, you can apply conditioner to keep the hooves supple.

★   ★   ★

Preventing Thrush

To avoid thrush (an infection in the clefts of the frog), clean the hooves daily with a hoof pick, never a sharp object. With the proper tool, you'll be less apt to injure the softer parts of the foot, or yourself, if you slip while digging out a packed hoof. Clean the whole foot from heel to toe, making sure you get all the debris out of the grooves next to the frog.

CHESTNUTS & ERGOTS

The chestnuts, or horny growths on the inside of a horse's front legs (a few inches above the knees), sometimes grow large and unsightly. You can periodically clip them with sharp hoof nippers if they don't shed or flake off on their own as they should. However, if you keep them from drying out by oiling them with mineral oil, baby oil, or petroleum jelly when you groom the horse, the older portions will usually fall off. Or if you periodically bathe your horse, the chestnuts may soften enough for you to peel off the outer layers.

A few horses have large ergots, horny growths that are usually hidden in the hairs at the back of the fetlock joints. Usually the simplest technique for removing them is to pull them off after they soften from bathing. Another method is to cut them with sharp nippers.

Conditioning Coronary Bands and Heels

Hand lotion will soften coronary bands and heels. Some of these creams aren't as expensive or greasy as hoof dressings, and they're nicer on your hands when you apply them.

Dressing Hooves

An easy way to apply liquid hoof dressing (especially if you make your own) is with a baby bottle and paintbrush. Enlarge the hole in the nipple just enough so you can pull the paintbrush handle up through it (the bristle end fits in the bottle). The handle should be snug enough to prevent leaking. Then when you unscrew the top, the paintbrush will be part of it, with its own catch pan to keep the hoof dressing from running down the handle as you apply it.

Letting Hoof Polish Dry

When using a commercial hoof polish, protect each freshly polished hoof from dirt and dust by placing it on a paper plate until it dries. You'll need to monitor the horse to be sure he doesn't move around.

Dressing the Coronary Band

Here's a simple way to keep a softening dressing (such as lanolin-based salve) on the coronary bands from collecting dirt in a paddock or brushing off as the horse walks. Cut a scrap of fleece into a strip big enough to encircle the coronary band, then punch holes in the strip about 2 to 3 inches apart and weave a shoestring through them. Rub the dressing onto the fleece and wrap the strip around the horse's pastern so that it rests on top of the coronary band. Tie the shoestring to hold the strip in place, making sure there are no dangling ends for the horse to step on with another foot.

NATURAL HOOF DRESSING

If you don't want to use a dye or commercial hoof polish on your horse's feet, you can still give them a little shine by rubbing them with the cut surface of half an onion.

FEET

BATHING

When bathing a horse for a show or another special event, it's important to consider such factors as water temperature, shampoo concentration, and effective drying methods.

★   ★   ★

Bath Basics

Many horsemen like to wash their horses before a show to get them perfectly clean, or after a ride to remove all the sweat and dirt. Still, lots of horses are never bathed and look every bit as good. It just takes a little more effort to brush all the dirt and dried sweat out of the hair. Too much bathing can deplete natural oils in the coat, and some shampoos and conditioners can irritate sensitive skin.

If you do bathe a horse, it helps to have a grooming mitt, sponges, towels, a sweat scraper, a hose with an adjustable nozzle, and a bucket. Before you start the bath, brush as much of the dirt out of the horse's coat as possible. Remove any tangles from the mane and tail before the bath, because they'll be even harder to deal with later.

Once you're ready, wet the horse completely and apply equine shampoo that you've diluted (see next page), using a rubber mitt to work it into a lather over his body. Don't use shampoo around the eyes, nostrils or mouth; a damp cloth is adequate. On a hot day you can use water from a garden hose, but on a cool day lukewarm or warm water is better. Never use hot water, which can scald the horse's skin. Also, don't bathe a horse in temperatures cooler than 60°F/16°C, or in a breeze that might chill him before he dries.

KEEPING WHITE SOCKS CLEAN

After bathing your horse for a show, protect his white socks by wrapping them with disposable baby diapers. The tape fasteners allow you to adjust the fit. Unlike leg wraps, these sock wraps don't need washing after each use.

When you finish lathering the coat, thoroughly rinse all suds out of the hair and off the skin to prevent irritation. If the coat feels at all slippery, it still has soap in it. It should feel squeaky clean, creating some friction against your hand. Use a sweat scraper (or the edge of your hand) to remove excess water. After a good rubdown with towels, the horse can dry in the sunshine.

Diluting Shampoo

Never use undiluted shampoo, which takes a lot of time and water to rinse completely out. It's important not to leave any residue to irritate the skin or cause a rash.

The usual dilution ratio is about 9 parts water to 1 part shampoo, but read the label in case your product recommends another ratio. An empty, well-rinsed dish soap bottle (or other squeeze bottle) makes a handy container for mixing and applying diluted shampoo. Just leave a bit of air space at the top of the bottle so that you can easily mix the water and shampoo by inverting the bottle several times. Don't shake the bottle; that just creates a lot of bubbles. If there's any diluted shampoo left over, label the bottle for future use.

Another way to make sure you're not using too much shampoo is to apply just one thin line of shampoo along the horse's topline, from the poll to the base of the tail. Follow with the hose to work the shampoo down the sides of the body, diluting the soap even more in the process.

STAYING DRY WHILE WASHING YOUR HORSE

Sometimes water will run over your hand and wrist and down your sleeve when you bathe a horse. To keep your arm from becoming soaked, cut the toe out of an old, thick, absorbent tube sock and slip the sock onto your wrist. To help keep your body dry, make a reusable protective apron from a large plastic garbage bag. Just cut a hole in the top for your head, and a hole on each side for your arms.

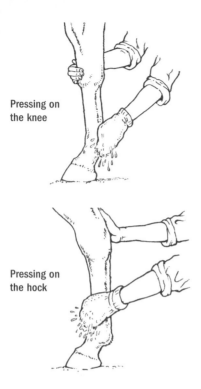

Pressing on
the knee

Pressing on
the hock

Washing the Legs

Before you wash a horse's legs, make sure he's comfortable about having them handled while you groom them. If not, postpone bathing them until he tolerates having them brushed or rubbed. When you do wash them, use a rubber mitt or grooming glove if they need to be scrubbed.

A leg is easier to bathe when a horse keeps his foot on the ground. If he tries to pick it up, encourage him to leave it down by pressing your free hand against the leg joint that he must bend to lift the leg. For a front leg, this means pressing against the front of the knee; for a hind leg, press against the back of his hock.

Drying the Tail

After washing and rinsing a tail, remove as much water from the hairs as you can with your hands, moving them down the tail with a squeezing (not a wringing) motion. Then squeeze the tail with towels, using as many as you need to absorb most of the water.

Spit Baths

If your horse needs a quick cleaning and it's too cold to get him all wet, try giving him a partial bath with hot towels. Fill a bucket with water that's too hot for comfort, and use rubber gloves so you can handle the hot, wet towels. Pour in a tiny bit of horse liniment or shampoo, then wet a towel in the solution and wring it out so it's just damp. Rub one part of the horse at a time until you've washed all the dirt or manure stain off, then dry that area with a clean towel. On a horse with sensitive skin, use just plain hot water, because you won't be rinsing him afterward.

CHAPTER SIX

Health & Medication

Here you'll find a wealth of tips for assessing
a horse's health and treating or dealing with
a wide variety of equine emergencies
and medical problems.

CHECKING VITAL SIGNS

A horse's vital signs (temperature, pulse, and respiration rates) and body language can give a good indication if a horse is healthy or ill.

★   ★   ★

Taking a Horse's Temperature

Normal temperature for a horse at rest ranges from 99 to 100.5°F/37.2 to 38.1°C. It's lowest on a cool morning, and slightly higher later in the day. It will rise above 102°F/39°C if he's been working hard and hasn't cooled out yet, or if he has a fever. You may want to check his temperature on occasion to see if he is ill or has been seriously overworked during a hard ride. It's best to use an animal thermometer because it's sturdy and has a ring on the end to tie a string onto, but any rectal thermometer will do. If you use a human thermometer, attach a string to the end with duct tape so you won't lose it in the horse's rectum.

Shake the thermometer down below 96°F/36°C and lubricate it with petroleum jelly or a bit of saliva in the palm of your hand, or immerse it in cold water, so it'll slide in easily without causing discomfort. Gently rub your free hand along the sides of the horse's tail; this feels good to the horse and he'll usually raise his tail instead of clamping it down. If he does clamp down, gently grasp the top of the tail head and pull it toward you, out of the way, as you slip the thermometer in. Don't try to raise the tail; doing so will alarm the horse and he'll try to clamp it down tighter. Aim the thermometer slightly upward as you insert it, so that it

---

DROP-PROOF THERMOMETER

To keep a thermometer from falling to the ground and breaking if the horse passes manure before you're through taking his temperature, do this: Attach an alligator or squeeze clip or a spring-type clothespin to the thermometer string, then clip it to the horse's tail.

---

will follow the contour of the rectum and not poke into the rectal wall. Spinning or rotating the thermometer a little as you push it in helps it slip in more easily so the horse hardly feels it. Let the tail go back to its normal position, then clip the end of the string to his tail to keep the thermometer from falling to the ground if the horse expels it. (see Drop-Proof Thermometer, previous page).

Leave the thermometer in for a full three minutes to ensure an accurate reading. If you don't want to leave it in that long because the horse is upset, one minute will give a reading that is 90 percent accurate (maybe only a tenth of a degree lower than his actual temperature), so a fever will still be obvious.

Taking a Horse's Pulse

The pulse rate for an average, healthy horse at rest is 24 to 40 beats per minute. Athletically fit horses generally have a lower resting pulse than fat, out-of-shape horses. It's easy to check the heart rate with a stethoscope placed at the girth, behind the horse's left elbow. If you don't have one, place the back or front of your hand firmly against the rib cage in that area. Remember that the actual heartbeat is two beats in one; each "lub-dub" counts as a single beat. You can count the rate for 15 seconds, then multiply by four, in case the horse doesn't want to stand still for a whole minute.

SELF-LUBRICATING THERMOMETER CASE

Petroleum jelly is a good lubricant for a rectal thermometer. An easy way to store a thermometer, if you don't have a plastic case, is to put it into a 20 cc syringe case half full of petroleum jelly. Poke a hole in the lid of the case just large enough for the thermometer to fit through. Wipe the thermometer completely clean after each use, then stick it back into its lubricating case.

Taking the pulse under the jaw

Another easy place to take the pulse is along the underside of the jaw. Run your fingers along the bottom of the bone until you find the artery (it feels like a small cord), then press lightly with your finger to feel the pulse. Other places you can find the pulse, once you figure out where to press, are on the temporal artery near the back corner of the eye above the cheekbone, and at the fetlock joint. There are also good arteries under the tail and just inside the forearm, but they are sometimes harder to find.

Checking a Horse's Respiration Rate

The respiration rate for a normal horse at rest is 8 to 12 breaths per minute, although in some individuals it's as low as 4 or as high as 20. You can easily check your horse's breathing rate by watching his nostrils or flanks move. When he inhales, his rib cage expands and his belly drops at the flanks. When he exhales, the flanks rise as he pushes the air out. Count the number of times he inhales or exhales, but not both.

Don't put your hand in front of his nostrils to feel his breath, or he'll try to smell your hand, which will alter his breathing rate. Neither should you put a hand on his flanks to feel the movement, as your touch will distract him. Just watch his breathing rhythm. Count for 15 seconds and multiply by four. If the horse is distracted and breathes erratically for a few seconds (some will hold their breath when listening or looking at something), do it again. Check several times to gain a true idea of his breathing rate.

DEHYDRATION PINCH TEST

When a horse has a fever, hasn't been drinking enough, or has become severely dehydrated during hard work or hot weather, his skin loses elasticity. Check his hydration status by pulling up a pinch of skin on his neck or shoulder to see how fast it sinks back into place. If it springs back immediately, he's not dehydrated; if it takes a few seconds, he is. For more information, see below.

Checking Capillary Refill Time

One way to check a horse for dehydration (when he is ill or has been working hard on a hot day), blood loss, or shock is to check his capillary refill time. In a healthy horse with good blood pressure the tiny blood vessels near the skin surface, or capillaries, are full and will quickly refill if you temporarily press the blood out of them with your fingertip. But if the horse's blood pressure is low (as it would be when he's dehydrated or in shock), his blood volume will be low and his capillary refill time slow.

The easiest place to check is in his mouth, on the gum above the incisors. Firmly press your thumb or fingertip long enough to leave a white spot in the pink tissue surface. After you remove your finger, count the seconds it takes for the white spot to turn pink again. In a healthy horse, the capillary refill time is between one and two seconds. Longer than two seconds sig- 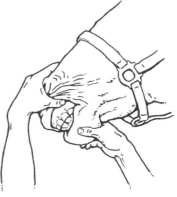 nifies a problem. If you're on a long ride, it means the horse is dehydrated. If his capillary refill time is more than five or six seconds, he may be going into shock. Another clue to his condition is the color of his gums, which should be bubble gum pink. If they're white, blue, or muddy red-purple, he needs veterinary help.

INTERPRETING A HORSE'S BODY LANGUAGE

Every horseman knows the signs of colic (pawing, rolling, and sweating), but mild abdominal pain may be harder to detect. The horse may be a little dull or restless or off his feed. He may get up and down more than usual, or spend more time lying down. He may lie with his nose tucked around toward his belly or stand off in a corner away from the herd.

If he's acting strangely or looking dull, check his vital signs, as an elevated pulse can be a sign of pain. Also check his abdominal sounds with a stethoscope, if you have one, or press your ear to his belly to listen for gut sounds. Remember, strange behavior or an abnormal posture may be a sign of something seriously wrong that a veterinarian should check.

Here's what a horse's posture may be telling you about his health.

★ Pointing one front leg forward or resting the same hind foot all the time may mean it's sore.

★ Standing with his hind legs rigid (often behind their normal position), and perhaps pawing the ground with his front feet, may mean he has muscle cramping in his hindquarters and doesn't want to move.

★ Standing with his front legs too far back and his hind legs forward (so that all four legs are bunched together) usually means his body hurts.

★ Shifting his weight onto his hind legs or placing his front legs out in front of his body in an attempt to take weight off his front feet usually means extreme pain (possibly founder) in the front feet.

ADMINISTERING MEDICATION

Giving your horse dewormer or other medications can be an easy matter if you use a few tried-and-true methods.

★   ★   ★

Disguising the Flavor of Medicine

Here are some tricks to get your horse to take his medication.

★ Most horses today are dewormed with paste preparations squirted into the back of the mouth. Some horses don't like anything placed in their mouths and they may flip their heads up and spill the medication, rear up, or rush backward. The best way to change their minds is to give them tasty treats for a while with an oral syringe, or a turkey baster with a rubber bulb (when giving larger quantities of medication). Try apple-sauce, pancake or corn syrup, molasses and brown sugar, yogurt, or even molasses-flavored water. Once your horse willingly accepts his "medicine" from an oral syringe, you can use it as an appetizer just before you deworm him. You can also smear molasses on the deworming syringe, or add a little molasses and water to liquid ivermectin.

★ Always sweeten a bad-tasting medication, such as phenyl-butazone, with molasses or syrup. Pills are cheaper than paste and can be crushed and mixed with a sweetener. Shake the syringe just before dosing your horse, to make sure the solid particles don't all sink to the bottom. If your horse tries to spit out a medication mix, stir it into applesauce to thicken it — a thicker mixture is harder to spit out than a runny one.

★ If your horse won't eat medication or deworming powder or granules in his grain, disguise the flavor with 2 tablespoons of molasses or several spoonfuls of powdered drink mix, mixed thoroughly into the grain. Stir the powder into some applesauce, if your horse is partial to it.

HOW TO GIVE
DEWORMERS & MEDICATION

Here are a few points to keep in mind when medicating a horse.

★ Dewormer is best administered at room temperature. Cold paste won't come out easily; warm paste may not stick to your horse's mouth, making it easier to spit out.

★ Before administering paste, make sure there is no grass or hay wadded up in his cheeks that will allow him to roll the paste and feed around in his mouth and spit it all out. Gently stick your finger into the corner of his mouth where there are no teeth, to encourage him to spit out any leftover feed. If necessary, flush it out with a syringe of warm water, repeating until his mouth is empty.

★ Because many medications break down when they're wet, always wait until you're ready to dose your horse before mixing the medicine with the treat.

★ Place the syringe into the corner of the mouth and deposit the medication as far back on the tongue as possible, where your horse can't taste it as well or spit it out. Tip his head up a little afterward, until he swallows the medicine.

★ Clean all syringes and mixing utensils thoroughly after each use so there is no residue left that can spoil.

Tip the horse's head back to ensure he swallows.

APPLESAUCE TO THE RESCUE

When traveling or camping with your horse, take along a few snack-size sealed containers of applesauce in your first-aid or camp kit. Then if you ever need to give your horse medication orally you can mix it with the applesauce and then add it to his grain or administer it orally with a syringe. The same goes for pills or boluses if you crush them first or dissolve them in a little warm water. The smaller the particles, the better they'll cling to grain and the more readily your horse will eat them.

Pulverizing Pills

A mortar and pestle are handy for crushing pills into a powder, but in a pinch, you can put a pill in a plastic bag and smash it with a hammer or other heavy object. A soft pill is easy to squash with a large metal spoon; a hard pill needs something that exerts more force, such as pliers. Yet another way to deal with pills is to blend them into a liquid with small amounts of water and molasses.

Figuring a Horse's Weight

Knowing a horse's weight is crucial for estimating the proper dosage of most medications and dewormers. Some people can guess the weight fairly accurately by visual observation, but many horses will fool you. A tall, lean horse may weigh less than a short, stocky one. The most accurate way to determine weight is with a livestock scale at an auction yard. Or you can go to a truck-weighing station and weigh your pulling vehicle with your trailer empty and again with your horse in it (just make sure he is the only change in cargo).

If there is neither type of scale nearby, use a weight tape to measure the girth and then read the estimated weight. You can usually obtain one for free at a feed store. If you can't find a weight tape, however, use a regular cloth or plastic-coated measuring tape or even a string that you can later measure with a carpenter's tape or a yardstick.

Consult the following chart for an estimate of your horse's weight, keeping in mind that it's based on an average body build.

GIRTH MEASUREMENT	ESTIMATED WEIGHT
62 inches	720 pounds
64 inches	790 pounds
66 inches	860 pounds
68 inches	930 pounds
70 inches	1,000 pounds
72 inches	1,070 pounds
74 inches	1,140 pounds
76 inches	1,210 pounds
78 inches	1,290 pounds
80 inches	1,370 pounds

More accurate than a weight tape is a formula based on girth and body length. Here's how it works.

1. Measure the girth in inches just behind the elbow, taking the reading right after the horse exhales.

2. Measure the body length in inches from the point of the shoulder straight to the point of the buttocks.

3. Multiply the girth by itself, then multiply that figure by the body length (heart girth X heart girth X body length).

4. Divide the total by 330 to find out the approximate weight of the horse. For example, if a horse's girth is 75 inches and his body length 64 inches, multiply 75 X 75 X 64 to arrive at 360,000, which you then divide by 330. The approximate weight is 1,091 pounds.

CLEVER WAYS *to*
REUSE PLASTIC BOTTLES

Recycled plastic bottles are handy for both storing and administering medicines. Here are a few ways to use them.

★ If your veterinarian dispenses pills or boluses in an envelope or small plastic bag, you may want to store them in empty, relabeled vitamin bottles, to protect them from moisture and make them easier to locate in your medicine kit.

★ When traveling or camping with your horse, plastic bottles of various sizes are great for carrying supplies such as sterile needles, tubes of ointment, and wound medication. They'll fit nicely in saddlebags, a glove compartment, or a trailer's first-aid kit. Other small items that are easy to lose, or hard to find when you need them, can go in plastic zipper bags.

★ Put wound powder, liquid wound dressing, or any antiseptic solution in plastic squeeze bottles, such as ones that contained ketchup or dish soap. Squeezing these items on a wound is much easier than trying to dab, pour, or spray them on. Squeeze bottles are also convenient for washing off a wound with warm water or a mix of water and hydrogen peroxide.

HOMEMADE LINIMENT

For a mild liniment that can help increase circulation, keep muscles from stiffening up after a hard workout, and ease a mild sprain or strain, mix equal parts rubbing alcohol and vinegar.

Applying Thick Ointments & Poultices

Certain kitchen utensils can come in handy for applying some topical medications. A serrated plastic knife (the disposable kind used at picnics) works well for spreading ointment or salve. It's flexible and the serrations help work the ointment down through the hair to the skin surface, plus it keeps the mess off your hands. A kitchen spatula is a great tool for spreading a poultice or other thick medication into all the nooks and crannies in the bottom of the foot. A spatula also makes it easy to scrape all the material out of the container.

Applying Liniment

A liniment relieves soreness and swelling in the muscles and joints by increasing circulation, thus removing waste products and fluids caused by inflammation. Avoid harsh liniments that can burn. A good rule of thumb is that if your skin can't tolerate the liniment, neither can your horse's.

A liniment will be more effective if you first hold a hot wet towel or sponge against the area for 10 minutes and then dry the skin before applying it. Sometimes a brace (a wrap on a lower leg) is used to enhance the action of a liniment. It shouldn't be too tight, however, or it may damage the tendons. Most liniments are best used after the initial inflammatory period has passed (24 to 48 hours); applied immediately after an injury, they will increase swelling.

DMSO and Nitrofurazone

DMSO (dimethyl sulfoxide) is an effective liniment for lameness and many leg problems. It can be used immediately after an injury because it works more like an ice pack, helping prevent swelling and tissue damage as well as increasing mobility of the joints.

Used alone, DMSO tends to burn and dry the skin. Mixed in equal parts with nitrofurazone ointment (antibiotic wound dressing), it is much milder and safe to use as a topical liniment or under a bandage.

HOUSEHOLD REMEDIES

Many human first-aid supplies and household products are effective on horses, which is handy when you don't have the veterinary equivalent. Here's a list of some familiar items and the ways to use them.

★   ★   ★

Fly Repellents & Insecticides

Surprisingly, these common household products have pest-fighting properties.

Petroleum jelly will kill botfly eggs. You can easily scrape away or pick off the eggs attached to the leg and chest hairs, but eggs laid between the hairs under the jaw are hard to see. Smear a little petroleum jelly there to smother the eggs and prevent them from hatching. You can spread it on your horse's jaw and legs daily during botfly season to keep the flies from laying their eggs on the hairs in the first place. Petroleum jelly mixed in equal parts with fly repellent can be applied around the eyes, ears, and nose to provide longer-lasting protection. Apply a small amount of jelly on the delicate insides of the ears to soothe existing fly bites and prevent additional ones.

Mineral oil mixed in equal parts with fly repellent will provide lasting protection from flies as well as prevent botfly eggs from sticking to the hairs. Use a cloth to apply the mixture to sensitive areas on the horse's face and to the edges and interior of the ears.

Avon Skin-So-Soft Bath Oil diluted with 3 times as much water and poured into a spray bottle makes a good fly spray. It's nontoxic and helps repel most insects. You can also moisten a cloth with it and wipe around the horse's ears and eyes, a mare's udder, or a male horse's sheath.

Vinegar diluted with 2 or 3 parts water and sprayed over the horse's body in a fine mist will repel most types of flies for short periods, depending on how much the horse sweats. If you wish, add a dash of Avon Skin-So-Soft, or a few drops of citronella oil. Apple cider vinegar added full strength to a horse's grain or drinking water also seems to offer protection. Flies still land on the horse but rarely bite. Perhaps some of the vinegar's properties are excreted through the horse's pores, or the pH of his skin is changed enough to discourage the flies. Start with a few drops and gradually increase it over several days to ⅛ to ¼ cup poured over the grain, or up to ½ cup added to 5 gallons of water. The idea is to use the minimum amount needed for it to work. If the horse doesn't like the flavor in his grain, add a little molasses to the vinegar to help disguise it.

Garlic fed to the horse (you can buy 5-pound tubes formulated for horses) may deter flies. There is no scientific evidence that it works, but horsemen who use it claim it does.

Pine-Sol mixed 2 or 3 parts to 1 part water, and with a few drops of tea tree oil, makes a good fly repellent. You can also add a few drops of citronella, eucalyptus, or pennyroyal oil to the mixture.

Listerine mouthwash combined in equal parts with baby oil makes an effective deterrent to some of the microscopic critters that cause dandruff and tail itching. Pour the mixture into a squirt bottle and apply it to the tail head, rubbing it into the roots of the hairs. The mouthwash kills the pests while the baby oil soothes the itch.

Hoof Dressings & Treatments
You don't have to go any farther than your medicine cabinet to find remedies for your horse's feet.

Petroleum jelly applied to the feet after washing can help keep the hooves from drying out. In winter, if you use it to coat the bottom of the hoof, it will keep snow from sticking and balling up under the foot.

Mineral oil helps protect the hoof wall from excessive moisture or drying. You can mix it with melted candles to make an all-purpose hoof dressing and polish (vary the proportions as desired to make it pasty or liquid). It can also be used on the bottom of the foot in winter to make a slick coating to prevent ice buildup in the hoof. Because it will not freeze, it works a little better than petroleum jelly for this.

Epsom salts (magnesium sulfate crystals) mixed in warm water are good for soaking an injured foot. A mixture of Epsom salts and tamed iodine can be used as a poultice to draw infection out of an abscess of the sole.

Sores & Wounds

A great many household products are useful in treating sore or wounded flesh on horses.

Hydrogen peroxide is great for washing and cleaning minor wounds. Its foaming bubbles penetrate the small nooks and crannies of a wound, lifting out dirt and contamination. Mixed with a veterinary antiseptic such as Nolvasan (chlorhexadine), hydrogen peroxide also makes an excellent flushing solution, taking the antiseptic into the depths of a dirty wound or abscess.

Alum is a very effective astringent for preventing proud flesh, the exuberant granulation tissue which often overgrows a leg wound and creates a mass of rubbery scar tissue. Available at grocery stores, alum is the white powder used in making pickles. Apply alum to a wound only after the new flesh has filled the gap; fresh wounds should be treated with something that disinfects the wound and protects it from further damage while stimulating the growth of new tissue. Traditional medications that prevent proud flesh are very caustic, but alum is less painful, more effective, and more economical. A small can is more than enough to treat a serious wound, even when applied daily for several weeks, because you only need to sprinkle a small amount over the surface of the healing wound.

Petroleum jelly can be used on a mild saddle sore, or on a rubbed area where the hair is gone, to keep it soft and protect it from fly bites while it heals. You can also protect hair and skin from the irritating effects of draining wounds or abscesses (or even a watering eye) by smearing the jelly over the area the discharge seeps onto.

Zinc oxide ointment is typically used on a baby's diaper rash, but it's especially good for treating raw or rubbed-away skin on horses.

Baby powder can protect and relieve minor saddle sores or irritations on a horse's back, as well as help prevent further chafing. Apply it to the area before you put on the saddle pad.

Aloe cream soothes saddle sores, cinch sores, or any other type of abrasion by keeping the tissues moist and acting as an anti-inflammatory. Spread it on minor wounds with a clean finger or a gauze pad.

Hemorrhoidal cream contains an anesthetic that soothes pain and itching, and medication to constrict small blood vessels and muscle fibers. It helps shrink a wound and stop the oozing of blood or serum, both of which attract flies.

Honey has some antiseptic qualities, and it's useful for soothing superficial wounds (but don't use it in deep wounds). For a wound dressing to protect raw scrapes from bacteria, mix 1 part honey with 2 parts fat (shortening, butter, or vegetable oil). Honey can also be mixed with petroleum jelly, zinc oxide ointment, or vitamin A&D ointment for a wound dressing that is especially beneficial for rope burns or extensive lacerations and abrasions where the top layers of skin are gone but there's no deep tissue damage.

Flaky Skin

These familiar products will alleviate some skin problems.

Mineral oil soothes and softens dry, flaky skin. Used on the tail area, it may halt rubbing due to skin irritation. A mixture of 3 parts mineral oil to 1 part tincture of iodine is very helpful in treating rainrot, a skin infection that creates tiny scabs; use it to work the crusty areas loose when grooming and bathing the horse.

Athlete's foot medication can be effective in treating mild fungal infections on a horse's skin. After washing the affected area with an iodine-based shampoo and rinsing thoroughly, apply the medication daily until new hair grows in.

Pain & Inflammation

Common aspirin can be administered to treat many of the same symptoms for which it is used in humans.

Aspirin relieves pain, inflammation, and fever. Even with newer anti-inflammatory drugs and painkillers available, it's still one of the safest and most effective medicines for horses, especially in treating leg problems. You can give the average adult horse from 7 to 45 grams of aspirin. The usual dosage for a 1,000-pound horse is fifty 5-grain (325 mg) human aspirin tablets, less for a smaller horse. You can also purchase larger animal boluses from your veterinarian. Crush the pills and mix them with molasses and milk of magnesia, to help buffer and minimize stomach irritation; administer the mixture by oral syringe.

Congestion

These everyday products can be useful in easing clogged airways.

Mentholated rub can help relieve congestion when a horse has a cough or difficulty breathing (for instance, from an allergy to dusty hay). The menthol helps liquefy mucus plugging the airways in the respiratory tract, making it easier to cough up. Antihistamines (which should only be used on the advice of a veterinarian) don't work as well in horses as they do in humans. They can actually hinder the horse's ability to remove mucus from the airways and may also cause excitement or depression in some horses, especially foals and older horses.

Dried mustard, horseradish, crushed fresh garlic, or chili pepper can be administered orally by syringe to relieve acute congestion. Several tablespoons of any of these items should be a sufficient dose. Mix them with water or some other liquid the horse likes to make them easier to administer.

FIRST AID

There are times you'll need to treat an injury or medical condition. You'll be able to act calmly and effectively if you already know what to do, or what to use for first-aid medications.

★   ★   ★

Treating Bruises

A bruise is a surface injury that doesn't break the skin but causes swelling due to bleeding under the skin and lymph seeping from the affected tissues. Applying cold water right after an injury slows the circulation and constricts the small blood vessels, which can help prevent or reduce swelling. Right after a leg or hoof has been injured, soak it in a rubber tub or bucket of cold or ice water, to reduce the pain in damaged tissue and help keep a knot from forming on a bruised bone.

> ## TRAVELING FIRST-AID BOX
>
> A portable first-aid kit stored in your trailer can come in handy when traveling to a show or trail ride. A toolbox or a fishing tackle box with drawers and compartments makes an ideal weatherproof container.

Treating Abrasions

A scrape may bleed or ooze because the top layer of skin is gone. The best treatment is to clean the area, then apply a soothing antibiotic ointment to prevent infection and crusting or cracking as the skin heals.

Cleaning a Dirty Wound

Because many antiseptics irritate damaged tissues, plain warm water is usually best for cleaning a wound. Use a washcloth, gauze sponges, or paper towels to gently scrub and remove hair, dirt, dried blood, or dead tissue until you arrive at clean, pink flesh. If the horse resents your efforts, you may have to resort to a twitch or Stableizer so he'll stand still.

GREEN WONDER

For a good antiseptic solution that cleans and flushes a dirty wound or abscess, mix ⅓ furacin solution (available only by prescription), ⅓ hydrogen peroxide, ⅙ water, and ⅙ Nolvasan (chlorhexadine). The antiseptic is called Green Wonder, a name coined by the veterinarians at Purdue University who came up with it. Just squirt the mixture into a puncture or lanced abscess with a large syringe — minus the needle, of course. The foaming, bubbling action of hydrogen peroxide takes the disinfectant deep into the wound or abscess.

Sometimes a horse will tolerate cold water hosed on a wound even if he won't let you scrub it; the water will dull the pain as well as loosen any dirt and dried blood. You can also use cold water in a well-washed squeeze bottle or a big syringe to flush out a small, deep wound. Once a wound is clean, you can evaluate it better and decide whether to call your veterinarian for more extensive treatment.

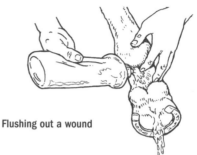

Flushing out a wound

Don't Overmedicate a Wound

When a horse suffers a wound, your first reaction is probably to treat it. But many horse owners overtreat wounds, sometimes doing more harm than good. After you've cleaned the wound, use a non-irritating wound cream or dressing to disinfect it. Never use iodine, methylene blue, alcohol, or caustic powders on a fresh wound; they burn tissues and destroy surface cells, which will delay healing and hinder the success of stitching the wound, should it need stitching. If you're not sure whether a wound needs sutures, just wash it or stop the bleeding and save any further doctoring for the veterinarian.

HANDY COLD PACKS

It's not always easy to keep an ice pack on your horse or to spend the time to continually hose a leg with cold water. A simple solution is to wet disposable diapers enough to saturate the absorbent material and freeze them individually in plastic bags. Wrap a frozen diaper around your horse's leg, and fasten it with its adhesive strips.

Another way to keep a leg cool is to soak a dish towel in water and wring it out. Fold it and place it in a plastic zipper bag, then lay it flat in your freezer for 30 minutes. When you take it out, let it thaw just enough to bend and conform to the horse's leg; bandage it in place. Replace it with another after it thaws. A package of frozen vegetables in a plastic bag will also work as a cold pack, especially if it's not a solid block and will conform to the leg so you can bandage it in place.

Sprains & Strains

A sprain is an injury to a ligament (connective tissue holding joints together), whereas a strain is an injury to a muscle or tendon (tissue that attaches muscles to bones). In severe sprains and strains, tissue is torn and may need corrective surgery. If your horse is very lame, make sure to have him checked by a vet. Many less severe injuries, however, will heal fine if the horse is laid off work and the injury treated with cold therapy for the first 24 to 48 hours (the sooner the better) to reduce the pain and swelling. After the swelling is gone (by the second or third day), heat and liniment can help, along with massage, to increase circulation. Don't use heat or liniment on a fresh injury, or you'll make it worse.

To cool an injured leg, hose it with cold water (below 50°F/10°C), or fill a soaking boot with ice water. Be sure the water is deep enough to cover the injured part, and add ice occasionally. Soak for a couple of hours at a time, or throughout the day if the injury is severe. Another way to keep the leg cold is to wrap it, then apply a second wrap loosely over the first one, slipping ice cubes between the two layers. If the horse doesn't improve by the second day, have your vet check for bone damage.

TIPS

WHEN IN DOUBT, CALL THE VET!

Your veterinarian is the best person to advise you on how to deal with a specific type of wound, as well as provide proper treatment to keep a horse from becoming unsound. If your horse is injured, consider the following.

★ Any cut over a joint or into a tendon needs professional attention, as does any deep cut below the fetlock joint. Once a joint is infected with bacteria, it may be permanently damaged. Harm to the coronary band may cause permanent changes in hoof growth.

★ Injuries in bony areas may need stitches, to minimize scar tissue.

★ Deep lacerations need to drain, and puncture wounds may have to be opened more fully and flushed out. Some wounds will need cleaning and medicating once or twice a day.

★ Treating a serious injury, such as a fracture, with anti-inflammatories before the vet arrives can make matters worse by diminishing the pain and causing the horse to move or even use the injured body part when he shouldn't.

Hot or Cold Therapy?

The important thing to remember is that cold water reduces blood circulation, and warm water increases it. In other words, if the injured area feels hot, use cold water; if it feels cold, use warm water. Hot or warm water (or liniment on a leg or muscle) is best when circulation needs to be increased. It can be beneficial for sore muscles, stiff and sore joints, or chronic low-grade tendon pain. A warm-water soak is also helpful on an infected area, such as a puncture wound, to help draw out the infection.

Fractures

Most fractures are the result of misstepping, falling, struggling after getting cast in a stall, or being kicked by another horse. If a horse won't put weight on a leg after an accident, consider the possibility of a fracture. Always play it safe and assume a fracture until proved otherwise, or you may jeopardize the horse's future. Don't move him or give him any medication until a veterinarian arrives him. Anti-inflammatory medication that reduces pain may mask the problem and encourage the horse to put weight on the leg, which could turn a simple fracture into a disaster. If he'll stand quietly, the best thing to do is just keep him still and wait for professional help.

If the leg must be supported, do it carefully. An improperly applied bandage or splint can worsen an injury, and trying to bandage a frantic horse can be dangerous. A fracture below the knee or hock will need a heavy support wrap, such as 10 or more quilted leg wraps under a firm bandage, or terry cloth towels or pillows; wrap the support with bandages to hold it in place.

No matter how high the break, start low on the leg and work up with the wrapping. The finished bandage should be at least 10 to 15 inches in diameter. You can try to stabilize the leg by taping 3-inch-wide boards of appropriate length to the outside of the bandage, not next to the horse's skin. Depending on the type of fracture, the horse may or may not have a chance for recovery, but often success or failure depends on the first-aid handling of the injury.

Pulling Porcupine Quills

Sometimes a curious horse gets too close to a porcupine and receives a noseful of quills. There are many old wives' tales about removing quills: soaking them in vinegar to make them come out more easily, twirling them as you pull them, or cutting them in two to let air out of the shaft (the idea being that a collapsed quill pulls out more readily). None of these are very useful.

The best way to remove quills is to first immobilize the horse. Have someone hold him still and distract him, or use a lip chain or Stableizer to keep him calm and relaxed (a twitch won't work, because his nose is full of quills). If you can't keep him still or he has a very large number of quills, he may need to be sedated — and you'll need your veterinarian for that.

In most cases, you can handle the problem on your own. Use needle-nose pliers to pull the quills out with quick, straight jerks. Don't pull to the side or the quills will break off, making them harder to extract. If there are a lot, you can sometimes remove three or four at once. Otherwise, it's best to pull quills individually; you'll have a straighter pull and be less apt to break them. Quills are easiest to remove soon after they are acquired; embedded quills tend to work their way deeper

HEALTH & MEDICATION

because of muscle action. Once all the quills are out, feel the skin to make sure none broke off below the surface. Feel inside the mouth and around the lips, too, to make sure none have worked through. Rubbing the nose will ease the sting where the quills have been, and your horse will appreciate it.

Stopping Sunburn

Horses with light (pink) skin or those with white face markings, particularly on the bridge of the nose, often sunburn. The burn is usually worst in areas where the hair is thin, providing little protection against the sun's rays. If your horse sunburns, you may be tempted to use a human sunscreen lotion. The various products are generally safe unless they contain PABA (para-aminobenzoic acid), which causes dermatitis in horses. If your horse reacts to a human sunscreen, wash it off and don't use it again. If necessary, use a steroid medication like dexamethasone to alleviate any swelling the sunscreen caused. To play it safer, use a zinc oxide ointment to block the ultraviolet rays. Another solution is to paint pink areas with an organic, nontoxic dye such as methylene blue or gentian violet; the dark color will stop the sun's rays.

Reducing Eye Glare

Horses with white faces, white around one or both eyes, or pink skin may be bothered by bright sunlight. The extra glare irritates the eyes and makes the horse more susceptible to cancer of the eyelid, because there's not enough skin pigment to protect the tissues from ultraviolet rays. The problem can be just as bad in winter, with sunlight reflecting off the snow.

If your horse squints in sunlight, protect his eyes by applying a dark, nontoxic substance around them to cut down on the glare and reflected UV rays. You can use dark mascara or cattle pinkeye spray. The purple-blue dye in the pinkeye medication won't hurt the eye and will stay dark for many days. Another product that works well is theatrical greasepaint, used by actors to reduce the glare of stage lights. The greasepaint is a cream and can be easily applied around the horse's eyes, especially if it is warmed first.

Getting Rid of Ear Ticks

Various types of small ticks sometimes work their way into a horse's ears. The larva of the spinous ear tick may burrow into the ear canal, where it will grow for several months. A horse with ear ticks will shake his head, rub his ears, and often have a droop-eared look. To eliminate ticks, put a cotton ball soaked with camphor into the ear. Alternatively, obtain an ointment containing an organophosphate (insecticide) in an oil base from your veterinarian and squeeze it deep into the ear with a rubber-bulb syringe. You may have to restrain your horse with a twitch or a Stableizer if he won't stand still for the treatment. Deworming the horse with ivermectin can also help control blood-sucking parasites, such as ear ticks.

Treating Fungal Infections

Most fungal skin infections, such as ringworm or girth itch, can be cleared up with a mixture of equal parts chlorine bleach and water; put it on the affected skin with a clean cloth or a squeeze bottle. If you do this once daily for about four or five days, it should be enough to clear up most fungal infections. Before using this treatment, however, dab a small amount of the mixture on the horse's skin and wait 24 hours to test his sensitivity to chlorine bleach. Some horses are more sensitive than others to certain medications.

Iodine is also good for treating ringworm, rainrot, and other fungal infections, but tincture of iodine is strong enough to burn a horse's skin. It should always be mixed 1 part iodine with 2 parts buffering ingredient like glycerin or mineral oil.

More Treatments for Girth Itch

Some horses develop raw areas at the girth, caused by a fungus. The skin may have been irritated or rubbed by a cinch, giving the fungus access. Even a tiny scuffed place can become an ugly sore. The infection can be halted most rapidly if treated early, before the horse has raw spots that make it painful to wear a saddle.

Wash the entire girth area twice a day with a good fungicide. Nolvasan (chlorhexadine) mixed with water (1 tablespoon in ½ gallon of water) is effective, as is tamed iodine. Another good remedy is Captan powder, a garden fungicide, mixed with water (follow the directions on the label) and used as a wash on the affected girth area. Yet another option is to thoroughly bathe and dry the skin, then apply an antifungal cream or ointment, such as Nolvasan ointment, or an antiseptic cream containing aloe. Work it in, leaving the skin soft but not gooey, so it won't collect dirt and invite further infection.

Once the fungus is halted, the sore should heal quickly. To prevent it from spreading, do not use the same cinch on other horses unless you've washed it in chlorine bleach first. If you think there's a chance another horse may develop the problem, wash his girth area with a dilute fungicide, such as ½ cup of bleach in 1 gallon of water, as a preventive.

TOUGHENING TENDER FEET

Some horses have thin, sensitive soles that bruise readily when ridden on rocky ground or gravel. Horses with flat soles also bruise easily unless shod with pads. Often you can get by without protective hoof pads, if you regularly apply tincture of iodine to the soles. This helps drive the quick, or sensitive flesh, farther back, toughening the feet. Iodine is also a good remedy for a horse that tends to be tenderfooted for a day or two after being shod.

Apply the iodine just on the sole, to avoid burning the skin or drying out the hoof wall. An easy way to prevent spills is to use a small syringe; slowly squirt the liquid a little at a time over the sole. Apply only enough to cover the sole without running under the shoe, where it can corrode and eventually weaken the horseshoe nails, perhaps resulting in lost shoes.

Applying iodine on the sole

BOOTS & BANDAGES

Here are some tips for treating feet and other injured areas, and for keeping them covered while they heal.

★   ★   ★

Protecting a Hole in the Sole

A hoof abscess (from a bruised sole or a puncture in the bottom of the foot) must be opened to flush it out and drain it, and then soaked daily for a few days until the infection clears up. Opening the abscess leaves a hole. To protect from contamination, plug the hole with soft dental wax, available from a dentist; it'll usually stay put even in water or mud. When the wax wears away, replace it until new sole material grows in.

Cleaning an Infected Foot

A rubber tub works well for soaking an infected foot because it's flexible and won't hurt the horse or make noise and scare him if he bumps it. Start by getting him used to standing with his foot in it. Rinse the tub and fill it with clean water (as hot as the horse can comfortably handle) and ½ cup of Epsom salts (magnesium sulfate crystals), then wash your horse's foot and place it in the filled tub. Soaking the foot for 20 minutes once a day for three to four days is usually enough to draw out the infection and let the puncture wound or abscess start to heal. Keep the foot clean and bandaged between soakings, using a hoof boot if necessary to keep the bandage from becoming wet or muddy.

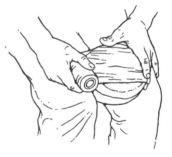

Bandage the foot between soakings.

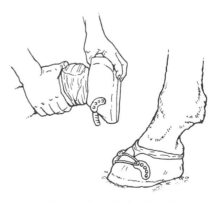

Use a hoof boot to keep the bandage clean.

Making a Soaking Boot

If you don't have a soaking boot to keep ice water or a poultice against your horse's foot or leg, you can make one from the inner tube of an old tire. Cut the tube so it becomes a long sleeve to pull over your horse's front leg; make it several inches longer than the length of the leg from the knee to the ground. Cut a few small slits in the rubber at the top end for threading a strap or baling twine to hold the sleeve in place. Also make a few slits in the bottom of the tubing, and thread another strap or length of baling twine through them. When the sleeve is on the horse's leg, fold the bottom back up against the leg and secure the strap around the pastern. Fill the tube from the top with soaking material, then either secure the strap around the horse's upper leg or attach it to a strap over the horse's back.

Thread straps or twine through holes in the top and the bottom of the tube.

Fill the tube from the top, then secure the upper strap.

PROTECTIVE BOOT FOR A PONY OR YEARLING

An Easyboot or some other type of hoof boot works great for an injured hoof or sole, but if you don't have one that's small enough, you can make do with the toe of an old overshoe. Just trim it to fit over the bottom of the horse's foot and up over the hoof wall a bit so that the edges can be secured to the hoof with duct tape. The tread on an overshoe is thick enough to protect the foot and provide traction.

Controlling Bleeding with a Pressure Bandage

You can usually stop a profusely bleeding wound by immediately apply-ing a pressure bandage directly over the spurting artery or flowing vein. Place clean towels, shirts, cloths, gauze sponges, or a similar material directly on the wound, and wrap it in place with a support bandage or leg wrap.

If a major blood vessel has been cut, you may have to make a more effective pressure bandage. Place a clean piece of wood or a smooth rock padded with bandaging material against the bleeding wound and secure with several layers of bandages. This will usually exert enough pressure against the vessel to halt blood flow. Don't make the bandage too tight or use a tourniquet; doing either will cut off circulation.

A horse's blood clots slowly, so don't be alarmed if blood seeps through the bandage for 10 to 15 minutes. As long as the flow has slowed, the horse will be all right until the veterinarian arrives. Don't leave a horse unattended after you apply a pressure bandage; monitor the bandage to be sure it stays in place and remains tight enough to halt bleeding, but not too tight in case the injured area swells.

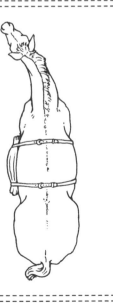

PUTTING A PRESSURE BANDAGE ON A HARD-TO-BANDAGE AREA

If the wound is on a part of the body that is not easily bandaged, put a clean towel over it and apply pres-sure with your hands. For a wound on the torso or barrel, fold a towel into several layers and hold it in place with blanket straps, bungee straps, or other straps long enough to go around the horse's body. If necessary, put a clean plastic garbage bag (folded to the proper size) between the wound and the towel to halt air flow. Continually check the bandage as you would any pressure bandage applied to control bleeding (see above).

Make-Do Bandages

You can fashion a perfectly adequate bandage from any number of materials found around the home.

★ Cloth diapers and old sheets torn into strips make good padding or bandage wraps.

★ Sanitary napkins provide excellent padding for small wounds; they're absorbent and won't stick to the wound. They're also handy as emergency pressure bandages to halt bleeding.

★ Disposable diapers can effectively draw moisture from a wound; the plastic on the exterior will protect the wound, and the self-sticking tabs will hold it on the horse's leg until you can cover it with a bandage wrap. Because they're durable and waterproof, disposable diapers also make excellent padded bandages for hooves. Boy's diapers are best because they're thicker in the middle, offering more padding for the foot. Reinforcing the outside edges and the bottom of the foot with duct tape will make the bandage last longer.

★ Bubble wrap makes a good temporary pad under a pressure bandage or a support bandage, and it has the advantage of distributing pressure. For example, you can apply it to the lower leg as first-aid care of an injury such as a bowed tendon; be sure to change it within three to six hours, because it will hold heat and moisture against the leg, possibly irritating the skin. With an open wound, first apply an absorbent layer (such as a sanitary napkin) and then the bubble wrap.

★ Nylon stockings, panty hose, or tights can be used to hold a bandage in place, if no commercial elastic bandage is available. They also make a good "mask" for facial wounds that have been sutured or stapled; just cut holes for the horse's ears and eyes, then stretch it over his head. To hold a bandage or poultice on the withers, sew two pairs of panty hose together at the waistband to form a rectangular area to go over your bandage; tie the panty hose legs together under the girth. Because the material stretches, it won't cut off circulation or chafe.

★ Terry cloth wristbands, which are stretchy and washable, are ideal for holding medication in place over a minor wound on the fetlock, pastern, or other hard-to-wrap place. You can buy them at most sporting goods stores.

Keeping Flies Away from Wounds

Flies chew on wounds, spread bacteria and infection, and lay their eggs in damaged tissue.

★ Apply zinc oxide ointment, such as Desitin, around the edges of an unbandaged wound. Flies don't like the smell.

★ To protect a leg wound, cut the feet off a pair of tube socks and slip the socks over the wound while it heals. You can spray a little fly repellent on the outside of the socks, if you wish. Socks will also help keep salve on a wound longer. Knee-high nylon stockings also work well because air is able to reach the healing wound but flies are not. A little fly repellent can be sprayed on the outside of the socks.

★ For a hock wound, use the elastic top from a stretchy, ribbed sock to hold the bandage in place; cut a 2-inch slit through which the point of the hock can protrude. Another option is to use a sock with a reinforced heel, placing the heel over the hock and taping the sock in place around the top.

★ To cover a knee, use a tube sock with the foot cut off; be sure to tape only the top, so the horse can still bend his knee without pulling the sock out of position.

★ For a wound around the heel, coronary band, or pastern, an old sock pulled up over the foot and kept in place with stretchy adhesive tape around the top of the sock will keep medication in and flies out. In some cases, just the stretchy top of a tube sock is adequate to keep medication on the heel or pastern; it will conform to the shape but will not be tight enough to hinder circulation.

HOW TO APPLY A BANDAGE

Some wounds heal better if left open. This is the case with most muscle wounds, as well as some wounds on the lower legs. Others should be covered to keep out dirt, support the edges of the wound, prevent stitches from pulling out, or keep medication on the wound. When in doubt, consult your veterinarian. If bandaging is required, follow these steps.

1. Clean the wound thoroughly.
2. Cover the wound with a nonstick dressing.
3. Apply several layers of sterile gauze pads.
4. Wrap the wound with stretchy bandaging material, to hold the pads in place.
5. Cover it all with sheet cotton or a light towel, then securely bandage it with a layer of cloth.
6. Replace the bandage every two or three days (or however often your veterinarian advises), or when it slips out of place, comes undone, or gets dirty.
7. If a horse chews on the bandage, smear the outside of it with something to discourage him, such as Tabasco sauce, a mixture of vegetable oil and cayenne pepper, or dish soap.

Applying an Eye Bandage

Early detection and treatment of an eye injury or infection may make the difference between recovery and blindness. Contact your veterinarian as soon as you discover the problem, and put the horse in a barn or shed (out of bright sunlight) until the vet arrives. If the horse is rubbing the eye, you may need a temporary bandage to protect it from further trauma and contamination. You can make a simple eye cover by attaching a folded towel to that side of the halter with large safety pins. If the horse tries to rub off the cover, tape a gauze bandage over the eye; hold it in place with an elastic bandage wrapped over the top of the head between the ears, over the eye, under the jaw, and back around to the injured side of his head.

Another way to protect the temporary bandage is to cover it with a panty hose mask. Cut off both legs of the panty hose to create a single large hole. Cut three holes in the seat area, just below the waistband, to accommodate the horse's ears and good eye, then roll up the seat to form a ring. Slip the ring, waist first, over the horse's muzzle to the bone bridge of his nose. Unroll the ring upward over his face, carefully adjusting the holes over his good eye and ears; cut a slit at the throatlatch so it fits him comfortably and doesn't obstruct his breathing. Put his halter on over the face mask, then secure the lower edge to the halter's noseband with duct tape.

PROTECTING A CHEST WOUND

To keep a horse from chewing at a wound on his chest, make a protective bib from a terry cloth dish towel. Hem the top and bottom edges of the towel, forming channels to thread baling twine through. Tie the top of the towel around the horse's neck, securing the twine to his mane to keep it from slipping forward when he lowers his head to eat. Tie the bottom twine around one front leg. The twine will hold the bib in place over the wound so the horse can't get at it.

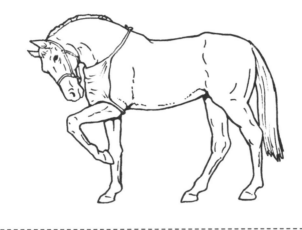

Putting on a Wet Dressing

If the eye is injured and the tissues raw, or if the horse has a head injury that needs protection until the veterinarian arrives, apply a wet dressing. Use a stack of gauze sponges soaked in a salt-water solution (1 teaspoonful salt in 2 cups distilled or boiled water), a thick sanitary napkin, or even folded washcloths or small towels dampened with commercial eye wash. The wet padding material should be moist but not dripping, and it should completely cover the injured eye or head laceration. Secure it in place with a bandage or a panty hose mask.

Makeshift Eye Patch

To protect an injured or diseased eye from dirt, sunshine, or flies, make an eye shield from the lower two-thirds of a lightweight, ribbed tank top and a soft, sturdy cup from a woman's bra. Pull the knit top over your horse's face until the top edge almost reaches his ears; mark the areas where you'll need to sew the bra cup in place to cover the bad eye, as well as where to

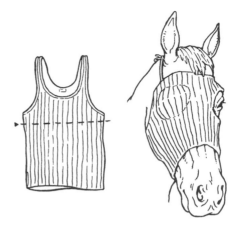

Trimmed tank top with bra-cup eye shield

cut a hole for his good eye. Remove the tank top to do this sewing and trimming. Sew two 15-inch-long ribbons or very wide shoestrings to each side at the top for tying the mask over the horse's poll, behind the ears. The stretchy material makes it easy to put the mask on and to roll or push it up to apply medication to the bad eye. You may want to make two masks so you'll have a spare when you wash one.

A simple eye protector can also be made from a bra (34C for an average-size horse), by cutting the center out of one of the cups for the good eye and fastening the bra straps to the halter.

Simple eye protector
fashioned from a bra

CHAPTER SEVEN

Foaling Season

Foaling time and the months spent raising
a new baby are filled with experiences that are special
and exciting. Here are tips and bits of advice on coping
with various situations that arise.

THE BIRTH

Most mares foal with no problems. If anything goes wrong, however, it's good to be there in case the mare or foal needs help.

★　　★　　★

When Will She Foal?

Some horsemen estimate gestation at 340 days (plus or minus 5 days), but others say 332 to 336 days (11 months). Some believe that colts (males) are carried longer than fillies (females). The fact is, mares generally don't foal on their due dates and, in some cases, have been known to foal as much as a month earlier or later. The following may help you guess when your mare will foal.

★ Regardless of age, mares foaling for the first time often have a shorter gestation period than veteran mothers. Mares foaling early in the year (January through March) tend to gestate an average of 10 days longer than mares foaling from April through June.

★ Old mares tend to carry foals a little longer than average.

★ Many mares have an individual pattern, always foaling early or late. Though some have a huge udder for weeks before foaling, others bag up overnight and surprise you with a foal the next morning.

Letting the Mare Take Her Time

After foaling, a mare usually lies there for 10 to 20 minutes, resting to regain her strength. The foal may still have his hind legs in the birth canal. Don't pull him out. This is nature's way of making sure he receives his full blood supply. The placenta contains blood that pumps into the foal for a few minutes as the uterus starts to contract. Other than making sure the sac is off his head and he's breathing, don't bother him or the mare. If the mare jumps up and breaks the cord too soon, her foal may be weak and anemic for a few days. Besides the possibility of the broken cord hemorrhaging, the placenta that was starting to detach and work its way through the birth canal may fall back into the uterus and become harder to expel.

Giving Artificial Respiration to a Newborn Foal

Occasionally a newborn fails to start breathing and, unless you take prompt action, may die. If the amniotic sac hasn't broken and the foal's head is enveloped in fluid, he will instinctively hold off breathing and may suffocate. Or he may be limp and unconscious after a difficult birth or if the umbilical cord breaks too soon and deprives him of oxygen. Whatever the reason, if he doesn't breathe within 15 to 20 seconds and has a heart rate lower than 90, he's in trouble. To check his heartbeat, place your hand on the lower left side of his rib cage, just behind and above his elbow.

To stimulate him to breathe, try tickling the inside of his nostril with a clean piece of hay or straw. If he doesn't respond (by coughing, sneezing, and taking a breath), give him artificial respiration. Here's how to do it.

1. Make sure his air passages are not obstructed. Roll him up onto his breastbone, with his head and chin resting on the ground, so his nose is as low as possible and fluid can drain out. Use a small suction bulb to clear his nostrils, if you have one with you. If not, press your thumb and forefinger along the top of his nostrils toward the muzzle and gently squeeze out any fluid as you would squeeze toothpaste from a tube.

2. Once the airways are clear, position the foal on his side, with his head and neck extended. Seal one nostril tightly with your hand, and hold his mouth shut. Blow gently but steadily into the other nostril until you see his chest rise. Allow the lungs to empty by themselves. Repeat this process until the foal starts breathing on his own or until the vet arrives.

3. Once he's breathing, prop him upright with bedding or hay bales until he can keep himself in that position. Being upright allows his lungs to fill more easily with air.

Disinfecting the Navel Stump

To minimize the risk of navel ill (joint ill), treat the navel stump with a good disinfectant, such as tincture of iodine, tamed iodine, or Nolvasan (chlorhexadine) immediately after the umbilical cord breaks, to keep bacteria from entering the foal's body. Horsemen have traditionally used tincture of iodine, but it can burn the foal's skin and some veterinarians advise against it.

Whatever you use, immerse the navel stump rather than swab it, to let the disinfectant seep between the layers of the broken umbilical cord and not just coat the outside. A small baby food jar or a shot glass works well as a container. As long as the stump is still moist, it's susceptible to infection. Disinfect it every few hours until it's completely dry; once it seals and falls off, bacteria can no longer enter.

Resisting the Temptation to Help

The foal may try to get on his feet and fall down several times before he becomes steady on his legs. You may be tempted to help him, but it's better to let him stand on his own and gain more strength and coordination with each try. For example, a foal that's had a difficult birth may have a cracked rib that will heal just fine because it's not separated or displaced. If you try to pick him up, however, you may displace the broken ends, which could then puncture a lung. Let him manage on his own unless he really needs help (if he doesn't stand up at all during the first two hours after birth), and then be very careful how you handle him.

Minimizing the Risk of Foal Rejection

On rare occasions, a mare will ignore, reject, or even attack her new foal, or seem confused and unsure about motherhood. Such a mare may have a hormone imbalance. Sometimes a physical problem, such as a sore udder, causes pain when the foal tries to nurse. But the all-too-common cause for foal rejection is human interference during the critical time when mare and foal should be bonding. A nervous first-time mother's maternal instincts may be disrupted by too much human handling at foaling time. Her instinct is to go off by herself, safe from interference by herd members. So stay out of sight while observing her, and don't rush

GIVING AN ENEMA

Many foals become constipated during the first 24 to 48 hours, having trouble passing the dark, hard balls of meconium that were in the intestine before birth. It may take one or more enemas to help the foal clear out this hard-packed material before the colostrum works through. If the foal is uncomfortable or straining a lot, he needs help.

The simplest way to administer an enema is to use a human adult enema product. These are sterile and premeasured. If you don't have one on hand, use a couple of drops of mild liquid soap (do not use antibacterial soap or strong detergent) or ½ cup mineral oil mixed with 1 or 2 cups of warm water. The warm water will soften the meconium, and the soap or mineral oil will lubricate it for easier passage.

While someone else holds the foal still, carefully insert an enema tube a few inches into his rectum. Be careful: The foal's rectum is delicate and you don't want to tear it. Slowly and gently squirt in the mixture. If you don't have an enema tube, squirt it into the rectal opening with the nozzle of a large syringe.

WHO'S IN CHARGE?

Horses establish a pecking order by either trying to thwart or reacting subordinately to the movements of their herd mates. If you control a foal's movement, you will impress upon him that you are the dominant one in his life. If you omit these lessons, he may tolerate your handling him but be disrespectful and disobedient.

in right after the birth, unless there's a problem and she needs help. Disinfect the foal's navel stump as soon as the cord breaks, then leave the pair alone.

Bonding takes place in the first hour, when the mare smells and licks her foal, identifying him as hers. Don't clean the stall right away, because the placenta and fluid-soaked bedding help the mare establish the smell bond that locks her newborn's identity into her brain. Drying the foal may also alter his smell and confuse her. Human interference can also distract the foal. His instinct is to follow any large moving object (under natural conditions, that's his mother), and he may be attracted to people in the stall or pen instead of seeking out the mare to nurse.

Imprinting

When first born, a foal does not fear people. He can be programmed to tolerate and remember many things. After that narrow window of opportunity, he becomes more like an adult horse in temperament (suspicious of changes in his environment and wary of new experiences or unfamiliar creatures that may be predators). By controlling what he sees and experiences right after birth, you can have a long-lasting, positive effect on his personality.

Because very young foals have fully developed senses and can soak up and remember a large amount of information, many horsemen like to imprint them early on so they'll accept rather than fear future encounters with people. Imprinting takes many sessions (the first time soon after he's born, even before he gets up to nurse) and involves handling the foal so he has the smell, touch, and appearance of humans stamped on his brain. It also introduces him to experiences he will be expected to tolerate later, such as someone touching his feet, ears, mouth, and under the tail.

If you intend to imprint the foal, keep his mother in mind. If she's had foals before and you know she will accept your handling him, fine. But if she's a first-time mother or very nervous, wait to make sure bonding has occurred before you do much with the baby.

Common Imprinting Mistakes

Unless imprinting is handled properly, it does more harm than good, because what the foal learns at this time is lasting. You must keep at each step until he relaxes and accepts it. A foal that's timid may still be apprehensive about what you're doing if you quit too soon. If the foal is bold and you halt the session before he completely submits, you'll merely reinforce his strong-willed determination to resist humans.

The most common mistake is to rush the first training session and make the foal more skittish instead of more comfortable with whatever you're trying to do, and more likely to resist in the future. The second most common mistake is to omit or inadequately perform the standing lessons. You should continue them periodically for a week or two, teaching the foal to stand patiently, lead willingly, and move forward, backward, and sideways on command.

NURSING

In most cases, a foal is on his feet and trying to nurse within an hour of birth. If he's unable to nurse, however, you'll have to help him.

★   ★   ★

Helping the Foal

A foal that has trouble getting his first meal (because he's weak from a hard birth, or big and clumsy) may need help to stand and suck or find the udder. Sometimes the mare is the problem, refusing to stand still, or kicking at him because her udder is sore. If he hasn't nursed by the time he's 2 hours old, help him, to make sure he receives the important colostrum. He needs a total of at least 16 ounces by the time he's 4 hours old.

With an uncooperative mare, put a halter on her and stand her against the fence or stall wall. If necessary, hold up one front leg to keep her from moving around to avoid the foal. This may be all the help a strong, vigorous foal needs to catch up with the udder. A weaker foal may need someone to guide him to the udder while you restrain the mare.

If your efforts fail, draw a little milk from the mare and offer it to the foal in a bottle to get him sucking the nipple. After he tastes the milk, he'll become more eager and you can then guide him to the udder with the bottle. This may be easier if you have someone on the other side of the mare holding the bottle underneath her and bringing it up next to her udder, to encourage the foal into moving in the proper position to nurse.

Milking a Mare

If all else fails, you'll have to continue milking the mare and bottlefeeding the foal. A mare will usually stand quietly for milking if the foal is next to her. If she's extremely uncooperative, you can use a twitch or Stableizer (see page 265) to make her stand still. Be sure your hands are very clean, and wet your fingers a little so they won't irritate the udder. Hold the container right under the teat so you don't lose any milk that squirts off to the side (the milk will come out in two streams from each nipple).

Using a Syringe to Milk a Mare

Because a mare's nipples are very short, they can be difficult to grasp with your fingers, making milking difficult. Additionally, some mares resent being milked. For those reasons, it's easier to use a human breast pump or a large (60 cc) syringe to draw the milk from the nipple. Either tool usually works well unless the mare won't let her milk down.

To adapt a syringe, cut off the needle end with a sharp knife or hacksaw, then rub it with sandpaper to remove any rough edges. After gently washing the nipple with warm water, place the smooth open end of the modified syringe over it. When you press it firmly against the udder and pull the plunger back, the vacuum created in the syringe will draw the milk from the nipple. The markings on the syringe will tell you exactly how much milk you've drawn (note that 30 cc equals 1 ounce). Empty the syringe into a nursing bottle, or into a container if the foal must be fed by stomach tube.

After using the syringe, take it apart and wash it in warm soapy water and rinse it in boiling water. After it's completely dry, put a little petroleum jelly on the plunger so it'll move freely next time. Store the whole thing in a plastic bag, to keep it clean for the next use.

Bottlefeeding or Syringe-Feeding a Foal

If the foal is able to nurse, you can use any clean, small-necked bottle (such as a soft drink or old lemon juice bottle) that a lamb nipple will fit onto. A lamb nipple is a better size for a foal than a calf nipple. A human baby bottle will work, too, as long as you enlarge the hole in the nipple a little for the milk to flow freely enough for the foal. Don't make it too large, however, or the milk will flow too swiftly and create the risk of it entering his windpipe. Whichever kind of bottle and nipple you use, be sure to wash them thoroughly between uses.

Hold the foal's head and neck between your arm and body so you can keep the nipple in his mouth as you teach him to suck the bottle. Once he figures out how to do it, and if he seems strong enough, you can try getting him to nurse the mare.

But if the foal is lying down and too weak to suck from a bottle, you may need to squirt the milk into his mouth with a syringe (one that holds 35 cc is about the right size). Sit on the ground next to him, supporting him with your body against his side to keep him upright on his chest and his head up and front legs forward. Never try to feed a foal that's lying flat; he won't be able to swallow properly and will inhale milk into his windpipe. Cradle his head and neck with your arm, to keep his neck extended and his mouth a little higher than his ears. When he's in this position, swallowing is easier because the milk runs slightly downhill.

While supporting his chin with your hand, place a finger in the front of his mouth to stimulate his sucking reflex. If he tries to suck, insert the milk-filled syringe alongside your finger and give him about 5 cc at a time, allowing him to swallow it before gently squirting in more. He'll need about 1 ounce of milk per 10 pounds of body weight per hour (for a 100-pound foal, that means 10 ounces of milk per hour), though you may want to give slightly more than that, to allow for spillage. For the first 24 to 48 hours plan on splitting the hourly total into three feedings, one every 20 minutes.

If the foal doesn't attempt to suck or won't swallow the milk, don't try to force him. Any foal that is unable or too weak to nurse should have veterinary attention as soon as possible.

Using a Mare Milker

A relatively new invention that makes milking a mare much easier is the Udderly EZ milker, a trigger-operated vacuum pump that snaps onto a flanged plastic cylinder attached to a collection bottle. The kit comes with a packet of sterile wipes (like those used on the teats of dairy cows), containing a disinfectant that is harmless to the foal and leaves no residue, taste, or odor. After wiping the teat, give it a squeeze to strip off any waxy material that may be plugging the end, and to avoid waxy flakes in the bottle. Next seat the flange of the milker over the teat and start pumping. After two or three pumps to create a vacuum, the milk will flow into the container and fill it very quickly. This device is easier on the mare than hand milking, during which friction from your fingers can make the teats sore. The pump is also safer, because you can reach below with one hand to use it without bending down under the mare.

Homemade Milk Replacer

If you're confronted with an orphan foal and have no substitute milk product on hand, you can make do with the following recipe, which approximates mare's milk. Genuine mare's milk is lower in fat and protein but higher in carbohydrates.

Start with a gallon of 2 percent (low-fat) cow's milk, which is a good base for the formula because it contains the proper calcium–phosphorus ratio for the foal. Add 1/2 cup honey or corn syrup (either is easier for a foal to digest than regular sugar) or 1/4 package of Sure-Jell (or any similar jam- and jelly-making product containing pectin) to increase the carbohydrate content of the milk. In addition to being a good carbohydrate source, pectin has a beneficial effect on the foal's digestive tract, helping prevent ulcers and diarrhea. If you use it, add a little honey to make it more palatable.

Dilute the milk mixture by adding about 5 cups of tap water (if yours has a high fluoride content, use distilled water). Divide the prepared formula into appropriate-sized portions for hourly feedings (about 2 cups each). Keep it refrigerated until feeding time, then warm it to body temperature, about 100°F/38°C, so it feels pleasantly warm but not hot to the touch.

MORE RECIPES FOR HOMEMADE MILK REPLACER

Here are some other recipe options for feeding foals.

OPTION 1
1 can evaporated cow's milk mixed with an equal amount of water
1 tablespoon corn syrup

OPTION 2
1 pint 2 percent cow's milk
1/2 cup (4 ounces) water
1 1/2 teaspoons corn syrup or honey

OPTION 3
(for several feedings)
3 pints 2 percent cow's milk
2 cups (16 ounces) water
1/4 package Sure-Jell pectin
1 teaspoon honey

FEEDING SCHEDULE FOR ORPHAN FOALS

A very young foal needs to be fed every 30 minutes for the first 24 hours. After that you should feed him every hour, starting with 1 cup of homemade milk replacer per feeding and gradually increasing the amount to 2 cups. By that time you will probably have located a good commercial milk replacer for foals, but you can raise a foal with the emergency homemade formula if necessary. Once he's 10 days old you can feed him at two-hour intervals during the night, but continue hourly feedings during the day until he's a little older and eating some solid food. As the length of time between feedings lengthens, be sure to increase the amount fed at each session.

Teaching a Foal to Drink from a Bucket

If the orphan foal is healthy and strong, you may decide to raise him on a bucket, which will be less labor intensive than feeding him from a bottle. To teach him, first get him to suck on your fingers and then immerse them in a bucket of warm milk as he sucks. If he has already nursed from a bottle, he may prefer the nipple and be reluctant. With patience, however, you can convince him to make the switch, especially if you put something sweet on your fingers (like sugar or honey) that he wants to suck. It also helps if you first use a shallow bowl instead of a bucket. When he dunks his nose into it, he'll touch the bottom without immersing his nostrils.

After he learns to drink from a bucket, you can use a commercial brand of milk replacer with the proper pH (such as Buckeye Mare's Milk) that will keep for several hours without souring. Hang a bucket containing several feedings of milk in his stall at a height he can reach, so he can help himself whenever he's hungry — and you can sleep through some of those night feedings! Always clean a bucket thoroughly between uses.

"Grafting" an Orphan onto a Nurse Mare

If you end up with an orphan, you might be able to locate another mare who will raise the foal. If her own foal died at birth, she will be more inclined to accept a substitute baby. The easiest time to persuade her is very soon after she herself has given birth, when her maternal instincts are strongest. Before you introduce them, liberally rub the mare's nose with Vicks Vapo-Rub (to mask other smells so she can't tell that the orphan is not her foal). Also rub her fresh afterbirth all over the orphan, especially over his rump and tail area, which the mare will nuzzle as he nurses. Some mares are so motherly and willing to accept a newcomer that disguising the smell is all you'll need to do.

If you're unsure about the mare's attitude or know that she has little tolerance for foals other than her own, you can increase your chance of success by tying the skin of her dead foal (basically just the back, rump and tail area) on the orphan for several hours until the bonding is complete. Before introducing the impostor, cover the mare's nose with the mentholated rub and hobble her hind legs so she can't kick the foal but can still move around. Choose hobbles that are lined with sheepskin or another soft material, in case you have to leave them on for a few days.

NURSING

THE IMPORTANCE OF COLOSTRUM

If the mare dies during or shortly after birth, make sure her foal receives colostrum (either milked from her or from a frozen emergency supply) so he will get the important antibodies he needs to protect him from disease. As an alternative, you can have your veterinarian give the foal plasma intravenously.

For the first nursing, have an experienced person hold the mare so she can still reach around and smell the foal, but can't hurt him or knock him down. If the orphan is desperately in need of milk, you can accomplish the first nursing quickly with the mare tied, hobbled, and twitched if necessary, and then let the bonding process take place afterward. Often a mare becomes more cooperative after a foal has nursed a few times, because nursing stimulates a hormone that makes her feel more motherly.

To make sure the mare doesn't hurt the foal, she should be restrained for each nursing and otherwise separated from him (in a divided stall where they can see and smell one another) until you are sure she accepts him. This may take one to three days, or even longer, depending on the mare. Some mares won't cooperate, and you may end up raising the baby on a bottle. But if you are diligent and lucky and can make the adoption work, the foal will be better off with a real mother. He'll grow up as a normal horse instead of thinking he's a human.

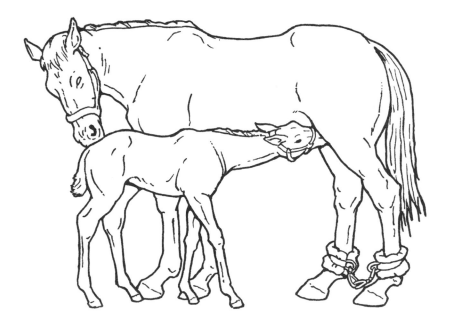

Restrain an uncooperative mare while the foal nurses.

Raising Two Foals on One Mare

Sometimes you can convince a mare to raise an orphan along with her own baby. This kind of arrangement usually takes more work and diligent monitoring for a while. Choose a gentle mare that gives lots of milk. Complete success is most likely if you can give her the extra baby immediately after she gives birth to her own. If you can present her with both foals at once (using a mentholated rub on her nose and smearing birth fluids over the orphan), she may think they are both hers.

Use adjacent stalls or pens so you can keep both babies away from the mare except at nursing time, and then restrain the mare before letting the foals in with her. Let them nurse at the same time, one on each side. When the foals are finished, return them to their pen. If you leave them with the mare, she will mother her own and reject the other. If you don't have someone to help with the foal shuttling, leave soft, close-fitting halters on them for easy handling. You'll also need a safe and manageable gate for moving them in and out.

After the mare is accustomed to the procedure, you can leave her free in her stall during the nursing session, just hobbling her hind legs so she won't try to kick the extra foal. Remove the hobbles once she accepts her role as mother of "twins."

BUDDIES FOR ORPHAN FOALS

An orphan will do better if he has a buddy: a pony, another orphan, or an older horse that tolerates him. The emotional problems for the foal are often a greater challenge than nutrition (today there are many good commercial feeds for foals). If an orphan foal has only humans for companionship, he may grow up spoiled. He needs to learn how to be a horse, and an equine companion will serve as a role model.

NURSING

Accomodating the Foals

Be sure to provide water for the foals while they're in their pen. They should also have access to feed (pellets for foals), although you'll probably have to put some of it in their mouths at first to help them learn to eat it. To keep them comfortable, the foals will also need good bedding in their pen (see pages 40 to 41).

Making Sure the Mare Has Enough Food

With the increased demand, a mare raising two foals will produce more milk and therefore need more feed and water. After the foals are a little older and the mare has bonded with them, they can be with her full-time at pasture. Separate the little family from other horses, and keep close watch on them to make sure all goes well.

CATCHING UP ON YOUR SLEEP

Because foals need to nurse often, you may become exhausted after a few days, especially when encouraging a nurse mare and foal or a single mare and two foals to bond. If you become desperate, you can leave a foal (or foals) with a still-reluctant mare to nurse at will for a while at night, as long as the mare is tied and her hind legs hobbled.

EARLY HANDLING

Here are some tips to make handling a foal easier, both when you're training him and when you're performing routine procedures.

★　　★　　★

Using the Mare to Corner the Foal

Catching a foal is easier when he and his mother are in a small pen or stall, where they don't have room to run. First catch the mare, luring her with grain if she's elusive. Tie her in a corner or have someone hold her so she can't move away or get between you and the foal, and so the foal can't duck under her neck to evade you. Now you can quietly corner the foal next to his mother.

Using an Arm Hold

It's better to initially restrain a foal with your arms rather than a halter on his head, because you're less apt to hurt him if he struggles. Put one arm around his chest or neck, and the other around his hindquarters. Block his movement (to keep him from going forward or backward), but don't put any pressure on him as long as he stands still. Just let your arms gently encircle him.

Do not attempt to hold him or pick him up with your arms around his rib cage; a foal's ribs are easily broken. Neither should you put pressure on or try to lift him by his abdomen, as that could injure him internally.

Using a Tail Hold to Restrain a Larger Foal

If a large foal is too strong and lively to restrain with your arms, you may have to use a tail hold. Place one arm around his front end at the base of his neck, and grasp the base of his tail with your other hand. If you can't reach around his whole body, grasp his mane just in front of the withers, to hold his front end. Lift the tail firmly but gently. With the tail straight up, most foals won't try to rush backward. Another means of holding a large foal is to pin him against a stall wall (or a solid, safe fence) with your hip pressed against his body while using the tail and mane hold.

Done correctly, tailing calms and immobilizes a foal, just as nose twitching immobilizes a horse. Endorphins released when pressure is applied to these areas are the body's natural narcotic and tranquilizer; they produce a feeling of well-being and increase tolerance of pain. Raising the tail up and forward also triggers the reflexes that straighten the hind legs, keeping the foal from kicking or struggling. When using a tail hold, don't bend the tail up too forcefully or you may injure it.

Turning an Adult Halter into a Foal "Handle"

A young foal is sometimes hard to hold on to when you're working alone and need to restrain him, lead him away from his mother, or lead him and his mother at the same time. Don't use a halter on his head; he may fight the restraint and even rear up and flip over backward, injuring himself. An arm hold is best, but may not work if you need one hand free for something else. To solve the problem, you can use an adult horse halter as a body harness. It lets you hang on to the foal with one hand and lead the mare with the other, or hold the foal with one hand while giving him an injection.

Use a full-size web horse halter with round buckles. Hold it upside down and slip the noseband over the foal's head like a collar, positioning it around the base of the neck above the shoulders so that the jaw strap rests on the withers. The crownpiece (which goes behind the ears of the adult horse) should be at his girth, where you can buckle it around the foal's belly like a cinch. Make sure it's not too tight, but it should be snug enough so he can't catch a hind foot in it.

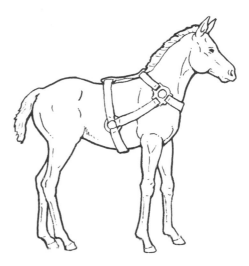

Halter foal "handle"

Using a Belt Hold

Another alternative for holding a foal when you're working alone and need to give an injection or some other treatment is to place a belt loosely around the base of the foal's neck. Once the belt is fastened, slide

your left arm (if you are right-handed) through the loop toward the foal's rear. Grasp the tail with your right hand and place it in your left hand. Thus your left arm and hand are holding the foal at both ends, with the help of the belt collar. With the foal in the belt hold and pressed between you and the stall wall, your other hand is free to administer the injection or treatment.

Belt hold

Teaching a Foal to Lead

The first few times you lead a very young foal should be with just your arms around his chest and his buttocks, to keep him under control as he accompanies his mother to and from a stall or pasture. After the foal is accustomed to handling, you can halter him. Gently corner him between you and the mare, or have someone restrain him with an arm hold so he can't back away or resist. Put the halter on carefully, without bumping his nose, muzzle, or ears

Haltering the foal

(especially if he was not haltered during early imprinting).

When you start leading him with a halter, don't hold on to him just with the halter rope; he may pull back or even throw himself down. To encourage him to go forward when he'd rather balk, loop a short soft rope or lead strap around his buttocks and hold the ends in your hand. Pulling on the rump rope will encourage him to move forward, away from the rope on his rear end. Pressure from the halter should come only if he tries to go too fast, never to pull him forward. Any restraint on his head may alarm him and cause him to rear up and possibly fall over backward, harming himself.

When using a rump rope, reward the foal for moving properly by releasing pressure. Pressure on the halter should cease as soon as he slows or gives to it, as should pressure on the

Leading with a rump rope

rump rope when he complies with a request to move forward. Keep the loop slack enough so he can move freely when asked but not so slack it hangs down and bumps his hocks. Once he leads well, you won't need the rump rope anymore.

Handling a Foal's Feet

When teaching a foal to have his feet handled, it's best to have one person hold the foal and another pick up his feet. Do not tie him. A foal may have trouble balancing on three legs; if he panics and pulls back on a tied halter rope, the lesson will turn into a bad experience for him.

Before picking up a foot, help him shift his weight so he can stand comfortably on three legs. Don't try to pick up the foot he has the most weight on. Instead, move him forward a step or reposition him. Run your hand gen-

Handling the foal's feet

tly down his leg. Don't just grab the foot or you may alarm him. Some foals will pick up the foot as soon as you touch that leg. Others refuse and must be coaxed, either by tickling the back of the heel or pastern or by gently squeezing the lower leg just above the fetlock joint. As the foal picks up his foot, lean into him a little to help him steady himself. It's easier for him to balance with a hind foot picked up than a front (unless you pull the hind leg too far up or away from him). The first few times, hold a hind foot quite low, just barely off the ground.

Don't let him pull the foot away from you, or he'll think he can whenever he wants. Keep holding it, and try to keep him balanced so he'll get over his panic. As soon as he relaxes, put the foot back down. If he succeeds in taking his foot away, pick it right back up again. Don't punish him, because he won't understand. Just be patient and persistent.

GROWING UP HEALTHY

Like adult horses, a foal occasionally needs medical attention. Here you'll find tips for administering and scheduling treatments that will keep him healthy and happy as he matures.

★   ★   ★

Dealing with Foal-Heat Diarrhea

One of the most common and least dangerous types of diarrhea in young foals occurs at the time of the dam's foal heat, which usually begins from 4 to 14 days after foaling, sometimes even later, and lasts up to 3 days. This is normal and does not mean the foal is sick. You can help him by applying mineral oil or petroleum jelly to his rear end to keep the acidic feces from burning his skin or causing hair loss. If his feces continue to be fluid for more than a day, or if he appears uncomfortable or colicky, you can give him about 30 cc Pepto-Bismol orally. It's better than Kaopectate, because it soothes the pain, slows down the diarrhea, has an antibiotic effect in the gut, and can be repeated as often as necessary. If you can relieve his stomach pain, a foal will often perk up enough to nurse and get the fluid he so badly needs.

Keeping a Foal Warm

Cold weather can be hard on a young foal, even if he and his mother are in a barn. One way to keep the youngster warm is with a zippered sweatshirt used as an overcoat. A child's sweatshirt will fit a small foal; an adult-size sweatshirt will work for a larger foal. Place the foal's front legs in the sleeves, and zip the sweatshirt under his tummy. If the sweatshirt has a hood, it can go over the top of the foal's neck.

Use a sweatshirt for a foal overcoat.

Most of the foal's body will be covered. A hoodless sweatshirt can be put on so that it zips along the foal's back. Another option is a dog blanket, which is less expensive than a foal blanket.

Giving a Foal Oral Medication

A disposable plastic syringe (minus the needle, of course) or a clean, well-washed deworming syringe works well for administering oral medicine. Gently hold the foal's head up so his mouth is tipped up a little and the fluid won't run back out. Stick the syringe into the corner of his mouth, aim it toward his throat, and push the plunger a little at a time, giving him time to swallow each squirt so he won't gag. Another option is a special dose syringe, which you can obtain from your veterinarian. It has a curved, stainless steel tube at the end that lets you stick it into the corner of the foal's mouth and deposit the fluid far enough back so he has to swallow it.

Treating Foal Ulcers

A foal that shows signs of abdominal discomfort (pawing, backing up, or even rolling and trying to lie on his back) may have ulcers. Administer 30 cc of Pepto-Bismol or a generic equivalent to coat the stomach lining and ease the pain.

Vaccinating a Foal

When the immunity that a foal derives from his mother's colostrum begins to wane after a few weeks, he'll need certain vaccinations. At 4 months old, he can start on some of his immunizations, for example, against Eastern and Western equine encephalomyelitis and West Nile virus. Depending on your locale and situation, he may also need to be vaccinated against other diseases, such as strangles, rhinopneumonitis, botulism, and rabies; consult your veterinarian. The influenza vaccination should wait, however, until he is older (8 or 9 months, if his mother was vaccinated), or he may not build adequate immunity.

The foal will also need booster shots, given several weeks apart. For most vaccines, he'll need at least three doses to start his immunity. Work with your veterinarian to figure out the most appropriate schedule.

Predicting a Foal's Full-Grown Size

A newborn foal grows swiftly. At birth he's about 10 percent of his mature weight and 60 percent of his mature height. His girth circumference is less than half the size it will be when he's an adult, and his lower legs are very long in proportion to the rest of his body. By the time he's a year old, however, he's reached about 50 percent of his mature weight, 90 percent of his mature height, and 80 percent of his mature girth size.

To predict how many hands a yearling will be when he's full grown, measure the youngster's legs after you get him to stand squarely on level ground. Hold one end of a string at the point of the elbow and pull the other end straight down to the fetlock. Keeping your hand at the elbow, bring the fetlock end of the string up above the withers, then add another inch to see how high his future withers will be. For a foal, add about 2 inches to the measurement.

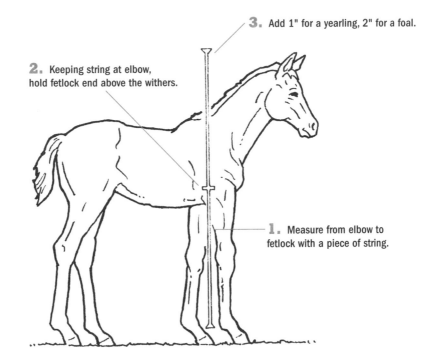

3. Add 1" for a yearling, 2" for a foal.

2. Keeping string at elbow, hold fetlock end above the withers.

1. Measure from elbow to fetlock with a piece of string.

WEANING

Weaning is always stressful for the mare and the foal, but this stress can be greatly minimized if you employ good management practices.

★   ★   ★

When to Wean

A foal can be weaned when he's 4 to 6 months old, or even later if the mare is not pregnant again. A 5- to 6-month-old foal is usually able to cope with both the physical and emotional stress of the separation better than a younger one.

PREPARING A FOAL FOR WEANING

A foal will handle weaning better if he's already accustomed to the feed he'll be eating. If he's only been on pasture with his mother, bring him in a few days before the actual weaning and feed him hay and grain. A foal will follow his mother's example and try the new feed. You can wean him after he starts eating adequate amounts of it.

Fence Line Weaning

The least stressful method of weaning is to put the mares and the foals in separate but adjacent pens that have safe, strong fencing that a foal won't try to jump and can't squeeze his head or a foot through. Small diamond-mesh or V-mesh wire fencing works well if it has a pole, pipe, or board on top so it can't be jumped or mashed down. Foals weaned in this manner usually spend most of their time near the fence and aren't worried. They still have their mothers' company, which is more important to them at this age than nursing. Within a week the mares will dry up and can be taken away. If the foals have each other or other horses for company, they should manage this transition smoothly.

Abrupt versus Gradual Separation

Gradual weaning (separating a mare and foal for increasingly longer periods for 5 to 10 day) is quite stressful, because it prolongs the ordeal and can cause udder problems in the mare and digestive problems in the foal. Abrupt weaning is easier on both of them. Leave the foal in a strong enclosure that's safe and familiar, and take the mare to a new place. She won't be stressed as much by a new environment, and she'll be better able to handle any new germs she's exposed to.

Weaning Foals in Groups

Traditional weaning involves putting several foals together in a pen and taking their mothers away. This is hard on foals because their desperation is contagious: If one runs up and down the fence, whinnying, they all do. They churn up dust (which is hard on their lungs) and rarely take time to rest or eat. They also take out their frustrations on one another, with the aggressive ones beating up the timid ones. It's actually less stressful for a foal to be put with an older "babysitter" horse that he knows (and will tolerate him) than with another foal. An old granny mare that likes foals is often a great help at weaning time.

Taking out a Few Mares at a Time

With a large group of mares and foals, one weaning method that works well is to periodically take a mare or two out of the pasture, far enough away so their foals can't hear them. Remove the mares with the oldest foals first. The newly weaned foals should stay relatively calm, because they'll still have their buddies and some adults in the group for security. By the time the last foals are weaned, the early weaned ones have adjusted and can provide security for the others. One drawback to this method is that sometimes foals are injured by mares left in the herd, when the forlorn foals pester them while searching for their mothers. If there's a particularly cranky mare in the herd, take her out first.

CHAPTER EIGHT

Trailering Horses

If you like to trail ride or compete in shows or other equine events, sooner or later you'll need to transport your horse somewhere. Here is a collection of trailer-loading and hauling tips to make your travel experiences safer and easier.

LOADING & UNLOADING

Getting in and out of a trailer is easier for the horse once he's done it a few times. But sometimes it takes patience and persistence to teach him that the trailer is safe.

★   ★   ★

Convincing a Hesitant Horse

Whenever you're faced with loading an inexperienced or reluctant horse into a trailer, use patience and common sense. Here's some advice that can be helpful in such situations.

★ Load a wary traveler with a buddy that is calm and trailer trained to serve as a role model. Load a foal with his mother.

★ Position the trailer so the sun shines into it; a horse doesn't want to step into the dark unknown. If he can see where he's going, he won't be as reluctant.

★ Give the horse a chance to smell the trailer, check it out, and relax. Don't try to lead him right up to it and make him enter, or he will balk. He must be convinced it's a good idea.

★ Convincing an inexperienced or nervous horse to step onto a trailer ramp can be difficult, because the ramp makes noise and doesn't feel solid. The horse will usually raise his head (a first reaction to anything unusual, for a better look at it), and this will make it even harder to get him into the trailer, particularly one with a low doorway. The first few times you load a horse, it's usually better to use a trailer that doesn't have a ramp. Most horses are not as reluctant to step onto the floor of the trailer (it feels more solid than a ramp). Because the horse will have to lift his foot 10 or 12 inches to step into the trailer, he'll lower his head at the same time to compensate, making it easier for him to enter without bumping his head.

★ For an inexperienced horse, position the trailer so the rear tires are in a low spot, with the trailer door near the ground. He'll be less afraid to step in if the floor is not so high. If he's still hesitant after checking out the trailer, and you lack the time for patient lessons, use a rump rope to encourage him to step forward. Do not try to pull him in; a horse will always pull back. Some slack in the lead rope, along with encouragement from behind with the rump rope or gentle tapping with a whip, always works better.

BACKING OUT WITHOUT A RAMP

When backing out of a trailer, a horse will typically remember the step up and do fine without a ramp. To make sure he never hurts a hind leg, weld a piece of rounded pipe or bolt some other smooth material along the underside of the back of the trailer, so there is no sharp edge against which he could scrape himself. If it's a large four-horse or stock trailer, you can let the horse turn around and come out front first.

Convincing a Reluctant Horse

One of the simplest, most stress-free ways to teach a horse to load is to feed him in the trailer for several days. It's easiest if you back up the trailer to his pen or stall gate and put his feed in the trailer, letting him go in and out as he wishes. If he hesitates at first, leave the feed just inside the trailer, so he can reach it while standing outside. Move it back into the trailer a little more each day until he is stepping inside to eat. When a horse doesn't feel pressured and entering the trailer is his own idea, he won't be as fearful or suspicious.

Feeding in the trailer

MAKING YOUR LOADING RAMP NONSLIP

A good way to improve traction on a trailer ramp is to cover it with an old trailer mat, discarded fire hose, or used conveyor belt (the latter is about 3 feet wide and can cover more ground than the fire hose). If you opt for the fire hose, cut it into lengths that will fit the ramp after you fold a little under at each end; secure the sections into place with wood screws and washers, making sure there are no uneven edges that could catch a shoe. Used conveyer belts can also be used as matting for stall or trailer floors, barn aisles, or even to create a wheelbarrow path to the manure pile.

BUNGEE-CORD DOOR HOLDER

To keep the wind from blowing a swing-style trailer door shut at an inopportune moment, use a bungee cord or motorcycle tie-down strap to hold it open. Hook one end to the side of the trailer and the other to the door.

Unloading a Hesitant Horse

It's always best to load and unload the horse several times when you first introduce him to a trailer, so he will be at ease going in and out. If for some reason a horse absolutely refuses to back out of a two-horse trailer, fasten a soft cotton rope to the rear of the trailer at shoulder height, and pass it around his chest and out the door on his other side. While verbally encouraging him in a soothing voice to back up, pull gently but steadily on the rope, providing an incentive to move backward (away from the pressure of the rope). Don't rush him, and don't allow the rope any slack or it will fall off his chest.

Using a chest rope

THE TRIP

There are a number of things you can do to make the trailer trip more pleasant and comfortable for the horse.

★   ★   ★

The Right Trailer Size

Make sure the trailer is big enough and tall enough for the horses you haul. A roof that's too low will inhibit a horse when he raises his head. A stall that's too short or narrow may hinder a large or tall horse in his ability to keep his balance by spreading his feet wider apart, and he may start scrambling or panic when you turn a corner. Many horses ride more comfortably in a four-horse or stock trailer with no stall dividers.

Facing Backward or Forward?

It's easier for horses to keep their balance in a trailer when facing backward. Horses are front heavy, and use their heads and necks to shift their weight and better keep their feet. Usually the most abrupt change in speed in highway driving occurs when you slow down. Horses quickly learn that when they're facing backward they can more easily keep their balance and compensate for the abrupt slowing so they don't fall forward. If they can't position themselves backward, they will turn at about a 45-degree angle, with their heads toward the rear.

Riding backward is usually impossible in a two-horse trailer; it puts too much weight on the back axle, and the trailer isn't designed for loading or tying horses that way. But facing the horses backward works well in a stock trailer. Many large horse trailers position the horses diagonally, with dividers to keep them in their slot, an arrangement that seems to work well, too.

Tying in a Trailer

If your horse's head is tied, make sure he has enough slack to use his head and neck for balance, but never so much that he can put a foot over the rope. Use quick-release snaps on the halter ropes, and always have a sharp knife handy in case you have to cut a rope in an emergency.

When using a two-horse trailer with a manger, be careful what you feed, because the horse can't put his head down to cough properly if there's a blockage in his esophagus. Good-quality grass hay is safer to feed than alfalfa, which contains lots of fine leaves that may collect in the manger. If a horse gobbles them down he may choke.

Roomier trailers like three- and four-horse trailers without mangers, and stock trailers with hay nets, in which horses can be tied with enough length to allow them to put their heads down to cough, are often safer and more comfortable than the confines of a two-horse trailer.

Hauling a Weanling or Yearling in a Two-Horse Trailer

When you must haul an inexperienced young horse that's not halter trained, it's safest to use a large stock trailer and just turn him loose inside. If all you have is a two-horse trailer with a divider, he may jump up into the manger or try to turn around and get into trouble. One way to keep him out of the manger is to fill the space completely with hay or straw. It'll also block the window and eliminate the temptation to get out.

A youngster that's small enough may attempt to turn around in the trailer stall. Prevent this by tying a rope to his halter and securing it to the front of the compartment at a height he can't lift a foot over. However, provide enough length so he can

PADDING YOUR TRAILER MATS

To give your horse an easier ride and prevent fatigue from standing on a hard floor, install a 2- to 3-inch foam rubber pad under the trailer mat. It'll help reduce road noise and heat, too.

THE TRIP

back up and touch the rump chain before feeling tension on the rope. That allows him to back up without having to pull and fight the rope, but keeps his head facing front.

INEXPENSIVE HOCK SOCKS

You can create hock padding from old tube socks. Cut the toes off four socks, then trim a couple of inches off two of the tubes and insert the longer tubes into the shorter ones. Fold the ends of the longer tubes over the shorter ones and stitch them together. Just slip these stretchy, double-layered socks up over the hocks.

Distributing the Weight

If you have a big trailer and several horses to haul, load the heaviest horses in the front to put more weight on the pulling vehicle rather than on the back end of the trailer. Otherwise the trailer won't pull properly. If there are just two horses, tie them toward the front, or use a divider to keep them in the front compartment if they're loose. With several loose horses, use the divider to keep some of them in the front and some in the back, so they won't all jam toward the back if you have to stop abruptly.

When hauling just one horse in a two-horse trailer, put him on the left side to compensate for the slope of the road (the center of the highway is usually a little higher than the outer edge), unless he prefers being on the right. Keeping the weight on the high side helps stabilize the trailer. Sometimes the outer edge of the road is rougher, so standing over the inner set of wheels means a smoother ride. On a long trip, one horse will be more comfortable if you take out the divider or use a divider that doesn't go clear to the floor. He'll be able to spread his legs wider for better balance.

To Wrap or Not to Wrap

Some horsemen use tail wraps (to protect the tail if the horse rubs it on the gate or back door of the trailer) and leg wraps while traveling. If you decide on wraps, learn how to apply them correctly. A wrap should be snug enough to stay in place, yet loose enough not to interfere with circulation. If a horse is unaccustomed to leg wraps, don't use them. If he paws and kicks because they annoy him, he may do more damage to his legs than if they weren't bandaged at all.

Allowing Extra Time and Being Prepared

When hauling horses, it's always safer to drive slower than the speed limit. For a trip that usually takes five hours in a car, allow at least six or seven hours with a trailer. You're always better off arriving at a show or trail ride a little early than having to hurry with a trailer, especially if you're going somewhere you haven't been before. To make sure of an early start, pack hay, grain, water buckets, feed tubs, and other supplies the day before, so you'll have little to load the morning of your trip besides your horse.

Final Check Before Leaving

Any time you pull the trailer, whether or not you have horses in it, first walk around the rig to make sure of the following.

- ★ No bicycles or kids' toys are lying behind or under the rig.

- ★ No one (at a show or other horse event) has set a tack box behind your outfit or tied horses too close to your rig.

- ★ All the door and tack room latches are fastened.

- ★ The trailer is properly secured to the truck, with the safety chains hooked.

- ★ The tires are properly inflated.

- ★ The wind-down jack (the one that holds up the trailer tongue when the trailer's unhitched) is up.

- ★ All lights and turn signals work.

Check Stops

Whenever you travel with horses, stop periodically to check on them and the trailer. If you have tandem axles, you may not be able to feel a flat tire until the second one blows out from carrying all the weight. Check each tire for heat (and missing chunks of rubber). If the tires are overheating, drive slower. Feel all the hubcaps (or the ends of the axles, if you don't have hubcaps) to make sure they're cool. A hub that's hot to the touch may indicate a dry wheel bearing or a rubbing brake.

When checking the horses, actually go into the trailer with them. Don't just glance in from the outside. If a horse is wearing a blanket, slide a hand under it to make sure he's comfortable, not cold and shivering or hot and sweating. A sweating horse may be sick or frightened by the ride. But if he's sweating because he's too hot, the trailer may not be well ventilated or his blanket may be too heavy.

Setting a Daily Mileage Limit

Try to plan a trip so you never have to go more than 500 miles per day. Some horses do fine in a comfortable trailer all day, but others fare better with a few stops where they can get out and stretch their legs. Some horses won't urinate in a trailer; others will, but only when it's stopped. Know your horses and accommodate them.

For a trip that takes more than a day, plan on a place to unload for the night where the horses can relax in a clean, safe stall or corral. If you must leave them in the trailer overnight, open all the vents and windows, and don't keep them in there for more than two nights in a row. A horse may develop digestive problems when confined in a trailer with limited ability to move around. Even though horses can doze standing up, they also need to be able to lie down sometimes for a more complete rest.

WHEN A HORSE WON'T URINATE

Most horses don't like to urinate on a solid floor because of the splashing. If that's true of your horse, try putting some sawdust or shavings on top of the trailer mat.

TIPS *for*
TRAVELING IN HOT WEATHER

Use a light-colored trailer, if at all feasible. It reflects sunlight and heat, whereas a dark color absorbs them. In fact, the inside of a dark trailer may be hotter than being outside in the sunshine. A horse confined for long in a hot trailer may overheat and become dehydrated from sweating or suffer heat stroke. To keep the trailer as comfortable as possible, do the following.

★ Plan ahead so you can leave early in the morning or at night when it's cool.

★ When traveling during the day, try to spend minimal time standing still once the horses are loaded; movement will generate a breeze through the trailer, which will take some of the heat with it as well as enhance sweat evaporation to help cool the horses.

★ Keep all vents and windows open for air flow. Remove the top doors or back curtains for more ventilation.

★ Don't blanket the horses.

★ In extreme heat, put ice blocks and wet shavings on the floor of the trailer, or pull into a gas station periodically and hose down the horses to help them keep cool. If they stop sweating for a while, they won't dehydrate as much.

★ Park in the shade, if possible. If you'll be stopped for a while, unload the horses.

HANDY PREPACKAGED RATIONS

When you plan to be on a trip, at a show, or on a ride for several days, mix the ingredients for your horse's total grain ration. Divide it into individual portions and put each in a plastic zipper bag so you won't have to contend with sacks of grain or have to mix dry supplements (you'll still have to give liquid ones). If you have more than one horse on the trip and they have different rations, label each bag with the horse's name.

Traveling in Cold Weather

Take along several blankets, including a well-insulated one to keep a horse warm and a lighter one in case he starts sweating under the insulated one. You want the sweat to evaporate so he won't stay wet and chill. Even in very cold weather don't close up the trailer completely; the horse needs some air circulation. He'll be better off wearing more blankets than riding in a completely sealed trailer. Moisture condensing on the inside of the windows means that the air is too damp, which can lead to respiratory problems or pneumonia.

Meals on Wheels

On a short trip it's better not to feed your horse. He'll be fine for two or three hours without feed or water, and there'll be less risk of colic, choking, inhaled dust from hay or grain, and other problems. On a long trip (more than four hours) the horse should be fed. The best thing to feed is clean grass hay (which is less dusty than alfalfa) or hay pellets. Hay is better than grain for minimizing the risk of colic if a horse doesn't consume enough roughage on an all-day trip. If you've been feeding him grain regularly, reduce his ration just before the trip and gradually increase it again afterward.

A horse eating hay will also drink more water — and you want to keep the horse drinking. Dehydration is one of the more serious problems that can occur when horses are traveling; lack of water interferes with proper gut motility, a problem that's compounded by lack of physical activity.

Feeding Hay in a Trailer

When feeding hay, it's usually safest to use a hay net and not a manger. Hay leaves that accumulate in a manger can cause choking and, with a manger in front of him, a horse can't put his head down to cough or drain off fluid from his blocked esophagus. If the fluid gets in his lungs, it can cause fatal aspiration pneumonia. But if you are using a manger, you could give the horse wetted hay cubes or a pelleted complete feed to reduce the risk of respiratory problems. Either will provide adequate nutrition without being as irritating to the air passages as dusty hay.

SWEETENING THE WATER

Many horses refuse to drink water away from home just because it tastes or smells different (due to a different mineral content, chlorine, or some other factor). Some horsemen bring water from home to resolve this problem, but if you are on a long trailer trip or plan to stay at a show or competitive event for several days, it's not feasible to carry that much water. A better strategy is to disguise the taste of the strange water.

It's easy to do this if you accustom your horse to a certain flavoring beforehand. A little molasses is a good bet. Other things that can be used to disguise the taste and smell of strange water include apple juice, apple cider vinegar, corn syrup, or a small amount of a powdered flavored drink mix. Oil of peppermint or wintergreen also works; one drop is usually enough for a bucket of water. Experiment with various flavors at home to see what your horse likes best or tolerates, and get him used to drinking it. You can bring a small bottle or package of the flavoring on your trip, instead of having to load gallons and gallons of water.

ARRIVING AT YOUR DESTINATION

When you unload a horse, walk him around for a few minutes to limber him up and accustom him to new surroundings. If there's green grass available, let him graze. A good appetite is a sign he feels well; a sick horse generally won't eat. Grazing will also cause him to lower his head, allowing any accumulated mucus to drain from his upper respiratory passages. And because green grass contains more moisture than hay or pellets, it will help him if he's a little dehydrated.

Whatever you feed a horse in the trailer or on the trip should be something he's already accustomed to; a change in feed may create more stress for his digestive tract. If you plan to use hay cubes or pellets, give him some with his regular feed for a few days before the trip.

Making Sure He Drinks

If the trip is longer than five or six hours, stop and water your horse at some point along the way. Some horses may be too nervous to eat or drink enough. Hot weather and fluid loss through sweating (compounded by the physical tension of the horse having to constantly balance himself) can lead to severe dehydration. Lack of fluid in the gut may lead to impaction, and low fluid levels in the body may cause muscle cramps.

To prevent dehydration, give your horse salt or electrolytes the day before you leave to encourage him to drink more water. You can sprinkle a couple of ounces of salt on his grain. If your horse won't drink while in the trailer, unload him periodically so he will drink.

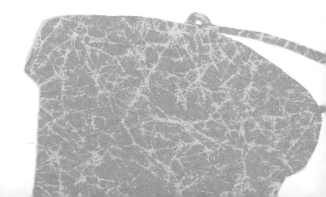

TRUCK & TRAILER MECHANICS

To haul horses safely, you need to know as much as possible about how your truck and trailer operate.

★　　★　　★

Making Sure Your Towing Vehicle Is Adequate

The larger and heavier the trailer, the more power and traction you need in the towing vehicle. The longer the wheelbase on the vehicle, the less it is affected by the weight behind it, and the easier it is to steer and keep straight while going down the road. The pulling vehicle also needs heavy-duty shock absorbers on the rear axle to take the extra weight and keep the vehicle riding level.

An automatic transmission generally gives the horses a smoother ride in the trailer during starts, stops, and changes in speed, but it may be inadequate in mountainous terrain. There you'll need to use the gears to help hold you back on downgrades, so you won't overheat and wear out the brakes and possibly lose them. Some newer pickups have automatic transmissions with grade retard. If you're buying a new truck, choose one with an adequate trailer-towing package.

Before you ever hook up a trailer, first be very familiar with how to drive your truck, whether it has an automatic or standard transmission. Be comfortable driving it on a city street, freeway, or backcountry road.

Trailer Checklist

If you pull your trailer a lot, conduct an inspection every few weeks. Regular maintenance is the cheapest insurance for trailering a horse safely. Here is a list of items to check.

Tires. Regular inspection is essential, because trailer tires can deteriorate even when you don't use the trailer. Place a jack solidly on the ground under one axle and raise the trailer until that tire is barely off the ground. Take off the hubcap and check the lug nuts to make sure they are tight; you don't want to risk losing a wheel. Also check that they are not rusted in place and impossible to remove if you ever have to change a tire. As you spin the tire, look for wear, nicks, and cuts in the rubber. There should be at least ¼ inch of tread with no sign of dry-rot cracks. Don't forget to check the condition of your spare tire, too. Find out how much air pressure your trailer tires should hold, if you don't already know, and take a tire gauge with you on every trip. Many flat tires result from improperly inflating the tires or overloading the trailer.

Wheel bearings and brakes. Wheel bearings that aren't regularly lubricated with grease may dry out and become hot because of the extra friction, causing the wheel to seize up or ruin the axle. Check them at least once a year or every 2,000 miles, whichever comes first. Keep a log of all your trips to tally your mileage. Check the bearings and brake shoes after a trip in which your trailer was subjected to salt air, dusty roads, sand, or gravel; any dust and dirt inside the brakes should be blown out with an air compressor. If you're not adept at checking and regreasing wheel bearings or checking the brakes yourself, let your trailer dealer's maintenance shop or local auto mechanic do it.

Checking for soft or rotten spots

Floors. Check the mats for wear. A solid but cushiony mat provides better footing, making travel easier on a horse, but if it starts to wear out, he may catch a foot in it. In a trailer with a wood floor, pull back the mats and use a screwdriver to check for any soft or rotten spots in the wood. If you have to replace the flooring, use pressure-treated wood. With a metal floor, look for signs of rust under the mats.

Interior. Look for bolts that may have worked loose and are sticking out. Inspect all welds between the door frames and trailer chassis and between wall sections; they should be solid and smooth, and all the joints sealed. Inspect wood walls closely, too; replace any damaged ones with exterior-grade plywood with the grain running horizontally. With vertical grain, a horse is more likely to pick up splinters when he strikes a hoof or leg against the wall. Tongue-and-groove boards running horizontally also make a good wall. For extra safety, put a good-quality rubber mat over the wood. After a period of nonuse, carefully check the trailer for any signs that bees, wasps, or ants may have moved in. And double-check to be sure there's no feed remaining from the last trip; you don't want any dust or mold to make your horse ill.

Exterior. Check all welded areas of the trailer frame for cracks, and make sure that none of the cross braces on the underside have come loose. Look for rust near the bottom and at the seams. On a trailer with a steel frame, pay particular attention to areas where the steel and aluminum sheathing meet, and make certain the rivets and bolts are intact. These are places where corrosion is likely to start. Look at the axles, too. Most have a slight upward curve; if they are straight or bowed down, they may be bent. Check all wires for loose or frayed connections.

Hinges and springs. Grease all the hinges and the springs (unless the trailer has torsion-bar suspension). Lubricate the door and ramp hinges with silicon ointment (available at any hardware or auto-supply store); it won't collect dust as oil will. If any hinge pins have shifted, tap them back into place with a hammer. Check the spring shackles (if your trailer has them) for wear.

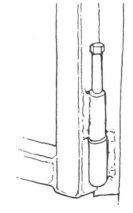

Check the hinge pins.

BUMPER PULL OR GOOSENECK?

Most small trailers have bumper-pull hitches, whereas most large or long trailers use a gooseneck hitch secured to the bed of the pulling vehicle. The gooseneck hitch puts the pivot point over the truck's rear axle and distributes the weight of the trailer on the entire frame of your pulling vehicle rather than just the back end. Using a bumper hitch with a heavy trailer puts too much weight on the rear tires of the truck, causing the front to stick up too high and making it difficult to drive.

Hitch. Keep the ball well greased to prevent wear. Check the brakes and the tail and turn-signal lights before each trip. Have someone stand behind the trailer to verify that they come on when you activate them. Replace burned-out bulbs or fuses, faulty or frayed wiring, and cracked or broken glass. Remember, you need a license plate light at night.

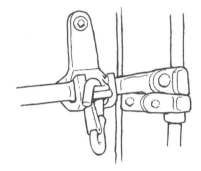

Secure a latch with a snap.

Latch. Make sure the trailer door has a dependable latch that will never open accidentally. Horses have fallen or jumped from trailers whose doors have swung open while the trailer was traveling down the highway. Some latch designs on new trailers are not dependable; they can pop open when you go over a bump. If your trailer has this type of latch, put a sturdy snap through it.

Windows. Check that all windows open and shut properly, and that none are loose, rattling, cracked, or broken. Replace any damaged windows with acrylic plastic, not glass.

EMERGENCY SUPPLIES & OTHER STAPLES

To be well equipped for an emergency or some other unforeseen event, stock your trailer tack compartment with the following items.

- ★ Fire extinguisher
- ★ Flashlight
- ★ Extra halter & lead rope
- ★ Tow rope for the vehicle
- ★ Jumper cables
- ★ Sharp knife
- ★ Hoof pick
- ★ Fly repellent
- ★ Scissors, twine & tape
- ★ Paper towels

- ★ First-aid kit (including sterile gauze bandages, stretchy bandages, topical wound medication, needles and syringes, petroleum jelly, squeeze bottle full of water or mild disinfectant solution; also a colic medication that your veterinarian recommends and shows you how to administer)

- ★ Tool kit (including a hammer, pliers, wire cutters, and a screw driver)

SAFETY CHAINS

Always use safety chains on a ball hitch, whether it's a bumper-pull or a goose-neck. If your hitch comes loose for some reason, the chains will keep the trailer from careening into oncoming traffic or off the road, and they can keep a gooseneck trailer from coming through the cab of your pickup. Some people like to cross the chains on a bumper-pull hitch; if the trailer tongue comes loose, the chain will cradle it and hold it off the ground. The chains should be strong and the proper length to keep them from binding on a corner or dragging on the ground, which will cause them to wear thin and break. Regularly check to make sure each link is sound; a chain is only as strong as its weakest link.

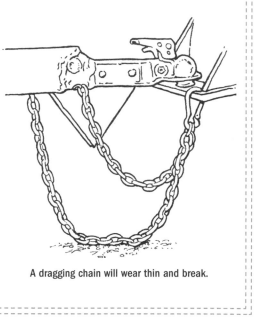

A dragging chain will wear thin and break.

Tips for Ball Hitches

Use the appropriate size ball to fit the trailer tongue; if it's too light or small for the hitch, the trailer may come loose while being towed. Keep the ball lubricated with a thin coating of grease. With a bumper-pull hitch, the height of the ball should match the trailer height so that the trailer is level when it's hitched up with horses inside. When the ball is too high (as it can be on some four-wheel-drive trucks), the trailer slants up in the front and nearly drags on the back end. That puts too much strain on the hitch (increasing the risk of breaking) and too much weight at the rear of the trailer. It's also uncomfortable for the horses, as they may be leaning on their rump chains or the trailer doors.

Make sure your hitch is strong. Factory-installed bumpers on older pickups are inadequate for pulling a horse or stock trailer. If you use a bumper-pull trailer (rather than a gooseneck), you may need to install a bolt-on or a custom-made hitch. It should fasten to the frame of the vehicle and not just the bumper.

Homemade Trailer Mud Flaps

You can fashion a multipurpose mud flap for a horse trailer from a section of heavy conveyor belt, which can often be obtained cheaply or free from a paper mill, lumber mill, stone quarry, gravel crusher, or other processing plant. It should be wide enough to cover the entire back end when attached lengthwise. Secure it in place with bolts and washers, allowing a 3- to 4-inch clearance between the belt and the ground to keep it from dragging. This type of flap helps prevent leg injuries when loading and unloading horses, because it stops a horse from putting a foot too close to the trailer. Not only does it help keep rain, snow, mud, and dirt from billowing into the back of the trailer when you're driving, but it also protects the vehicles behind you better than conventional mud flaps.

Hitch Protector

When your trailer is parked in the barnyard, pasture, or anywhere a horse may go by it, cover the hitch with a plastic bucket, an old bell boot, or some other soft, brightly colored object to keep the horse from running into it and injuring himself.

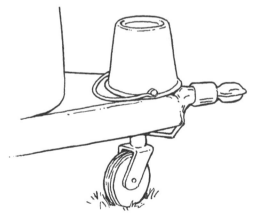

Protect the trailer hitch with a plastic bucket while parked.

ADJUSTING THE BRAKES FOR ROAD CONDITIONS

When driving on slippery roads, you want the trailer brakes to barely grab but not lock. You still want them to engage slightly ahead of the truck brakes, but be careful not to set them too high or they'll seize and jackknife the trailer.

Adjusting the Trailer Brakes

Most newer trailers have electronic rather than hydraulic brakes. You should always be aware of the adjustment on the electronic control unit. It's not something you can set when you buy the trailer from the factory and then forget about. You must reset it for your load. If you're pulling an empty trailer and the brakes are set for pulling horses, you'll drag the tires when you apply the brakes, possibly resulting in a skid, jackknife, or other kind of wreck. And if the brakes are set for an empty trailer when you're hauling horses, your truck brakes will do all the work and you won't be able to slow down enough to make a turn.

Electronic brakes are controlled from a unit (usually a black plastic box) that's installed under the dashboard of your pulling vehicle near the steering column where the driver can reach it. Adjusting the brakes involves two components: The first is a lever on the side of the box that should hang straight down; the second is a knob or thumb wheel on the front of the box, which enables you to fine-tune the brakes for changes in road and load conditions. The primary adjustment, done with the lever, sets the brakes in relation to your pulling vehicle when it is hooked up to the trailer; it sets the position of the pendulum inside the unit (which swings toward a magnet to make an electrical contact when you put your foot on the brakes). This initial adjustment is done before you hook up the trailer brakes. You can set it and then not worry about it.

Because the lever has to be set perpendicular to the ground, before you fine-tune the brakes, be sure the rig is on a flat surface. Then push the brake pedal by hand until the power indicator on the unit comes on. With your other hand, move the lever forward until the light goes from bright to dim. At that point the pendulum is set correctly. To be safe, do it several times to make sure it always comes to the same position.

Now hook up your trailer brakes so you can make the second adjustment using the thumb knob on the front of the brake unit. While pulling the trailer, coast at about 5 mph on level ground. Then with the knob adjustment set low, press your brake pedal. Slowly turn the knob until you can feel the trailer brakes gently engage, but not up so much that they lock or drag the tires. Once the brakes are adjusted, they should work fine until there's a change in your load or the grade. For instance, if you're approaching a steep downhill grade while pulling a load of horses, you can adjust the brakes to a higher setting before you head down the hill.

BEING PREPARED TO CHANGE A FLAT

Keep tire-changing equipment in the trailer at all times. You'll need a jack sturdy enough to lift your loaded trailer (or a trailer block to drive the wheel onto) and a large socket wrench for loosening and tightening lug nuts. It's wise to have two correctly sized and properly inflated spare tires mounted on the correct rims. (Bring along the tire and rim specs just in case you ever have to make an emergency purchase on a trip.) Don't forget to bring wheel chocks for blocking your wheels, if you ever have to unhitch the trailer from your towing vehicle.

Tire-Changing Ramps for Tandem Axles

If you don't have a heavy-duty jack for a tandem-axle trailer, you can make an inexpensive, stair-style tire block/ramp from wood. Start with a 2-foot-long 2 x 8 for the base, nail a 1½-foot-long 2 x 6 on top of it, and finish with a 1-foot-long 2 x 6 on top; make sure the boards are flush on one end and staggered on the other. Insert the ramp under the good tire and move the trailer forward or backward so that tire is on top of the boards. This will raise the flat tire off the ground and make it easy to remove and replace.

For a similar tire ramp, use a 2½ to 3-foot-long section of a 6 x 6 or beam. Saw off one end at a slant so you can pull the trailer up onto it, leaving the flat tire off the ground. You can position the ramp with the slanted side frontward or backward as needed for changing a tire on the front or rear axle.

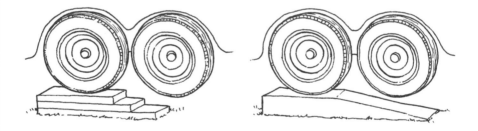

CLEANING TRAILER STALL MATS

Remove all the manure from the mat grooves, using a wire brush if necessary. If mats become really dirty, take them out and completely wash them and the trailer floor with a garden hose. Let the floor and mats dry completely before replacing the mats.

HINTS *for*
KEEPING THE TRAILER IN GOOD SHAPE

★ When you return home from a trip, clean everything out, including all feed, manure, and bedding. Remove the mats to allow the floorboards to dry. Sweep the floor, and scrub it with soap and water if needed.

★ Periodically use a wood preservative, such as log oil, on the floorboards (when the mats are out) so they'll last longer. If the floor is aluminum, check it for soundness, making sure there are no deep areas of corrosion.

★ Wash soiled trailer walls with dish soap, using a plastic scrubber to dislodge dirt and dried manure. Wash the exterior of the trailer several times a year, to remove mud and winter salt or sand. A coat of wax afterward will help protect the paint.

★ Metal trailers rust in spots where the paint chips off, so maintaining the paint not only protects the metal but also helps the trailer's appearance. Sand and repaint any rust spots, bare places, or scratches.

★ If possible, keep the trailer in a shed or under a tarp when not in use; this can add years to its life.

★ When unhooking a gooseneck, be sure your pickup tailgate is down when you drive away from the trailer so the hitch won't hit the tailgate and bend or destroy it.

DRIVING A TRAILER

Experience is the best teacher for pulling a trailer, but here are some useful tips.

★   ★   ★

Trial Run

If you've never driven a truck/trailer combination, schedule some practice sessions without the horses, especially if it's a big stock trailer or a four-horse trailer. Start off in a level field or a big parking lot, just to get a feel for how the trailer pulls and cuts into the corners. Spend some time making right turns. Set out cones or other visible obstacles and practice driving around them (without hitting them), backing up between them, turning corners while backing up, pulling into a tight "parking space" and so on. In this way, you'll learn how much space you need before you swing out into an intersection, back up between two trailers in a crowded lot or fairground, or park on a street. Remember, you can't turn as sharply while backing up with a bumper pull as you can with a gooseneck. Every truck/trailer combination is unique. Not only does each trailer pull differently, but having an extended cab or a short wheelbase makes a difference in the turning angle.

After you feel comfortable driving in the field or parking lot, spend some time maneuvering and parking in traffic before you load horses in the trailer. Drive downtown in the early morning when parking lots are relatively empty. Take someone with you who has experience pulling a trailer to provide tips and advice or help rescue you from a jam. It's also beneficial to have another pair of eyes to spot vehicles coming alongside you, assess how close you are to an obstacle when turning or parking, and read a map or road signs.

The first time you pull the trailer loaded, try to take horses that are already experienced travelers. You won't have to worry as much, because they'll know how to stand up in a trailer even if you take a turn a little too fast or sharply or make an awkward start or stop.

TIPS *for*
DRIVING SAFELY

Follow this advice to protect yourself, your horses, and other drivers on the road.

★ Always start slowly and accelerate gradually, for a smooth rather than a jerky takeoff.

★ Drive at or below the posted speed limit, slower if the horse is a nervous traveler or if the road or weather conditions are bad.

★ Avoid sudden swerves or lane changes.

★ Leave lots of space between your truck and the vehicle ahead, so you'll have enough room to stop without throwing the horses forward.

★ Take curves and corners slowly, and do not speed up again until the trailer is completely around the corner.

★ Reduce speed for bumps, even more than you normally would, so you won't jostle the horses. Make sure the trailer is past the bump before you accelerate.

★ Learn where your front and back trailer tires track in relation to your truck tires; you'll need to know for navigating tight corners or gates. Their tracks are usually a little wider than those of your truck tires (narrow two-horse trailers are an exception), so be sure to allow extra room for the trailer.

DETERMINING THE STOPPING DISTANCE OF YOUR LOADED RIG

Because all truck/trailer combinations vary a little, the stopping distance depends on such factors as the weight you're carrying, your speed, and the condition of your brakes. There's no magic formula for gauging the distance. The best way to figure it out is by driving your rig (with the horses on board) at various speeds along a road with very little traffic and making controlled stops. Pick a landmark to begin your stop and note the mileage; at the point where you come to a complete halt, note the mileage again. The difference between the two is your stopping distance.

Using the Side Mirrors

Set the mirrors each time you pull the trailer to adjust them so you can see as much as possible (there are always some blind spots). You need special mirrors that extend out farther from your truck than regular side mirrors, to give you a better view of your whole trailer. If you haven't driven a truck and trailer before, have someone walk around the rig at different distances while you check the mirrors to locate the blind spots, both alongside and behind your trailer. Check the mirrors often as you're driving, especially when turning or backing up.

Adjust the mirrors so you can see the spot where your tires meet the highway. The only clue you may have that a tire is losing air is that it's starting to bag out a little on the bottom; check often enough and you may spot the problem. Sometimes a tire will become hot and suddenly disintegrate, and you'll see the ragged edge flopping. If your mirrors aren't adjusted properly, you could go several miles with a destroyed tire, and the one next to it could also be ruined before you notice.

Preparing to Stop

The extra weight of the trailer and horses makes it harder to stop. Learn the stopping distance of your rig. The larger the rig and the more horses being hauled, the greater the effect on acceleration and stopping distances. Allow extra space, even when the trailer is empty. If you typically allow 300 yards to slow down in your pickup, for example, double that distance. With horses aboard, allow even more room.

Once you know your stopping distance, you'll know what you can and cannot do. If you're going too fast when you reach an intersection where you need to turn, it's better to go on by and come back rather than stop too quickly and possibly tip over the horses or jackknife the trailer. With a 20-foot horse trailer and four horses, for example, you have to allow as much as a third of a mile to slow down from a speed of 50 mph to negotiate a turn at about 5 to 10 mph.

Making Turns

Remember, your trailer will always make a sharper turn than your truck. The longer your trailer, the more its back end will move inside the radius of your truck, cutting the corner. A gooseneck will cut the corner even more tightly than a bumper pull. Swing wide whenever you turn, or the trailer may run over the curb or hit a stop sign or parked car.

Always allow extra room when making a 90 degree right turn at an intersection, but do not swing out into the left lane, even though this seems the logical thing to do. It's illegal as well as dangerous. Remember, you have a blind spot somewhere between your mirror and the back of your trailer. If you don't see a car that's trying to pass you on the right as you swing left, you'll run into it when you make the turn. The correct way to make the turn is to stay in your own lane until you actually round the corner, and then swing out into the oncoming traffic lane on the road you are turning onto. The oncoming drivers will see you making the turn and slow down to let you complete the swing and move back into your own lane. This is the only legal way for a big truck/trailer rig to make a turn at an intersection.

Slow down before a turn and around curves, so that your horse can keep his balance and not be thrown forward. It's easier for him to brace himself if there is less force throwing him to the outside of the corner. Take right-angle turns at 5 to 10 mph, and even more slowly when hauling inexperienced horses. Be sure the trailer is all the way around the corner before you begin to accelerate.

Accelerating

The heavier and longer the rig, the more time it takes to get up to traveling speed. Keep this in mind when accelerating or merging with traffic. You'll need to use all your gears (unless it's an automatic transmission) and run up to your engine's maximum torque rpm (revolutions per minute) on a diesel truck, for instance, before you shift. If your normal merging space without a trailer requires about 100 yards, for instance, allow at least four times that distance when pulling a loaded trailer.

Coping with Trailer Sway

Because most trailers have stabilizers, sway is rarely a serious problem, even when the horses are moving around. But if your load shifts suddenly, you may have a hard time controlling the rig. When the trailer exerts too much of a rocking side pull on a vehicle, there's even a risk of rolling over.

Don't hit the brakes. Instead, take your foot off the gas and try to keep the truck straight as you coast to a slower speed. Don't try to steer your way out of the sway as you coast. If you overcompensate for the trailer movement, the rig may jackknife or fishtail, possibly pulling you off the road. And hitting the brakes when the trailer is fishtailing will only make the problem worse.

Keeping Your Distance

Always allow plenty of space between your rig and the vehicle in front, especially when pulling a heavy load. That way, if you have to stop, you

★

can do it gradually. In town, a safe rule of thumb is to keep at least four car lengths between you and the next vehicle, especially when approaching an intersection. As your speed increases, your spacing also needs to increase.

On a freeway, make sure there's at least 100 yards between you and the vehicle in front. This gives you time either to change lanes or to stop if that vehicle slows or stops. On a road with several lanes of traffic, drive in the lane with the vehicles traveling closest to your speed, so you won't have to change lanes often or make other drivers pass you. Check your mirrors regularly, because traffic on both sides of you means more blind spots. When you exit, move into the turning lane well in advance so you won't have a problem finding room at the last minute.

Changing Lanes and Passing

Plan ahead when moving from one lane to another or passing a vehicle. Don't just look in your mirrors once and go. Check several times to make sure another vehicle has not come up behind or beside you from a blind spot. Use your turn signal well in advance, to give other drivers adequate warning. Lightly tap your brakes a couple times to alert the vehicle behind you that you're about to do something. The gentle braking will also warn the horses, waking them if they've been napping, so they can brace themselves for a stop or a turn. Preparing the horses this way is especially important after a long, straight stretch.

On a mountainous two-lane road, you may be traveling slower than the other traffic. Be aware of cars behind you; let them pass when possible. If you have to pass a vehicle, make sure there's plenty of room to drive around it and no oncoming cars; swing back into your own lane only after the trailer has cleared the vehicle. When judging distances, however, keep in mind that the mirrors in newer trucks make objects look farther away than they really are.

You'll encounter many tricky situations when pulling a trailer in traffic. Keep a cool head. Even if another driver becomes angry or impatient with you, don't let that color your judgment, cause you to rush, or make a decision that could endanger your horses. Safety is your prime concern.

Backing up with a Trailer

Backing up always takes more room than pulling forward. The basic principles are the same with a bumper pull and a gooseneck, except that a gooseneck trailer can make tighter corners without landing in trouble. Also, with a bumper pull you must be careful not to bind the bumper against the tongue.

When backing and turning, steer the pickup in the opposite direction you want the trailer to go. As the trailer begins to come around, straighten the pickup tires by turning the steering wheel back a little toward the center. Now turn the pickup tires the same way the trailer is going to follow the trailer around as it turns. Make small (not large) adjustments in steering as needed, especially with a gooseneck trailer. The trailer will react rapidly if you make too much of a correction.

Where space is tight, as when backing through a gate or into a tight parking space, you'll have to pull forward and start over if the trailer becomes too crooked. As you pull forward, turn the nose of the pickup in the opposite direction from which you started, to straighten it out. Watch the trailer in your mirrors and when it's lined up with the slot, try backing up again.

To back through a gate that's at a right angle to your vehicle, it's helpful to pull away from the opening to where you can see the gate from your window as you back up. Don't turn to the offside of the opening, or none of your mirrors will show you where your boundaries are. Your goal is to swing your vehicle around to where you can always see the gate as you approach, either from the driver's-side window or mirror. Never rely on your passenger's-side mirror. If you maneuver into a position where you can't see where you are backing, stop and pull forward a few feet (it doesn't take much) to straighten out into a position where you can maintain a good view.

In a four-wheel-drive truck with high and low range, put the vehicle into four-wheel low for better control when backing around a corner. Even people who pull trailers all the time usually shift down to maintain precision on turns when backing up. You are more likely to do it properly the first time if you are going slow enough to make small corrections as needed.

CHAPTER NINE

Knots & Ties

Knowing how to tie knots that won't work loose,
as well as ones that untie easily even when the horse has
pulled on the rope, is a valuable skill. This chapter
features a variety of useful knots and tips for
attaching ropes and tying horses.

HANDY KNOTS, TIES & HITCHES

Here are the knots that every horseman should learn how to tie. You never know when or where they'll come in handy.

★ ★ ★

Square Knot

If tied correctly, the square knot is the perfect knot for tying a broken rope back together or for tying a gate closed, because it will not slip. To tie a square knot, start with an overhand knot (see right). Next tie another overhand knot on top of it, but this time in reverse. Before you pull the rope tight, you'll see that the square knot looks like two linked, closed loops leading in opposite directions.

KNOT TERMINOLOGY

When you're explaining how to tie a certain knot (or figuring out how to tie a knot someone is describing to you), it helps to know the following: The end of the rope you're tying is the working end, and the remainder of the rope is the standing part. The simplest knot of all, and sometimes the first step in forming more complex knots, is the overhand knot. It's the knot you make first when tying your shoes, before you form the bow.

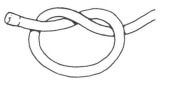

Manger Tie

Also known as a reefer's knot or a bowknot, this quick-release knot is good for tying your horse to a fence post or rail or to a ring in a manger. Like a square knot it won't slip, but it has the extra advantage of being easier to untie when it's been tightened, for example, by a horse pulling on the rope.

1. Start with the working end of the rope coming up and over the rail or pole you're tying to.

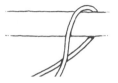

2. Bring the working end under the standing part (coming away from the pole) and double it into a loop.

3. Bring that loop over and through the first loop.

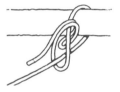

4. Pull until the knot is tightened.

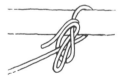

5. To untie the knot, just pull on the free end of the rope extending from the knot.

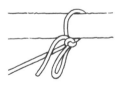

Bowline Knot

Probably the best nonslip knot for tying a rope around a horse's neck, the bowline won't tighten up when the rope is pulled and it's fairly easy to untie.

1. Hold the rope with the long standing end to your left, and the working end to your right. Form a right-hand loop by passing the working end over the standing part, and pinching the place where the rope crosses between your left thumb and index finger.

2. Insert the working end of the rope through the loop from the back, then wrap it around the back of the standing part from left to right. Reinsert the working end through the loop from the front.

3. Hold the working end of the rope and the right side of the loop in your right hand and the standing part of the rope in your left and pull to shape and secure the knot.

Clove Hitch

This simple wrap is a quick way to secure a rope around a pole, but it's not a quick-release knot. In fact, it actually tightens when pulled hard, making it harder to untie. Still, it can be very handy used in conjunction with a quick-release knot like a bowline, because it'll keep the rope from sliding along a pole or pipe.

1. Run the working end of the rope once around the pole and then over the standing part to loop around the pole again.

2. Poke the working end up under itself as you complete the second loop.

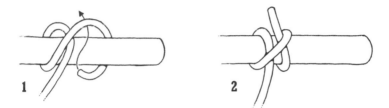

Note: *To combine this hitch with a quick-release knot, such as the bowline, first tie the simple clove hitch around the pole to keep the rope from sliding. Take the two ends that are left (the working end and the part that goes to the horse) and tie a bowline with them.*

Quick-Release Clove Hitch

In this version of the clove hitch, a quick pull on the rope will untie the knot. First make the simple clove hitch, then bend the working end and feed it back through the hitch. Tug on that end to release the rope.

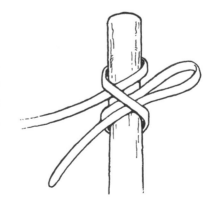

Ring Knot

Here's a good knot for tying a horse to a ring of any kind. It doesn't tighten up much when the horse pulls on it, and it only takes a quick jerk of the hand for the knot to undo and the rope to slide right out of the ring.

1. Fold the rope in the middle and twist it a few times, then feed the twisted part through the ring.

2. Pinch the middle of the standing end to form a small loop and feed it through the loop at the end of the twisted section.

3. Make a loop in the working end of the rope, and put it through the loop in the standing end, leaving a tail that can be pulled to untie the knot.

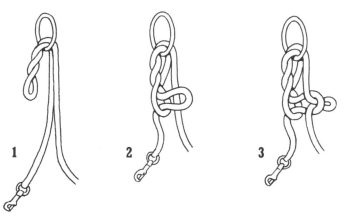

Quick-Release Tie

A properly tied release knot allows you to pull the end of the rope to undo the knot and quickly release the horse, no matter how tight the knot was pulled.

1. Put a loop of the halter rope through a tie ring or around a post (or whatever you're tying to). Then twist the loop a couple of times (this will keep the knot from getting pulled so tight that it will be hard to undo).

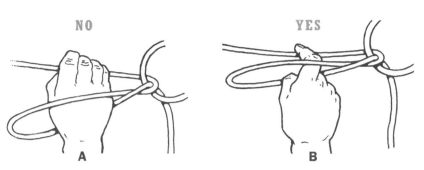

Never put your hand through a loop in the rope (A), instead use just a finger (B)

2. Make a loop in the standing end (leading to the horse) and bring it through the twisted loop. Do this with just one or two fingertips, to avoid having your hand caught in the loop if the horse steps back. Never put your hand or any part of your body through a loop in a halter rope.

3. Put one or two fingers through that second loop to grab a piece of the rope that is attached to the post or tie ring (standing end). Pull it through the loop, and tighten it to produce the final knot.

Weaver's Knot (Fish Line Knot)

A square knot may not hold for tying together two ropes that are very slippery or different in size. In those instances, a fish line knot (like you'd use to tie a leader to a fish line) works better.

1. Make a bend in the larger-diameter rope and thread the smaller rope through and then around it, coming back out under itself

2. Pull both ends tight.

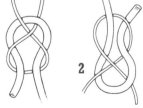

Note: *To tie another version of this knot, make a bend in the larger rope, thread the smaller one through it, around one side, under the other side, and then back over itself, tucking it back under the bend of the larger rope.*

Double Half Hitch

This knot is quick and easy to tie, acts like a slipknot, and is a handy way to secure a rope to something in a hurry when no other type of knot seems appropriate.

1. Position the standing end to the left of the post (or whatever you're tying the rope to). Take the working end in your right hand and wrap it around the post, passing it over and then under the standing part and up through the loop you just made around the post.

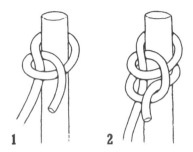

2. Repeat step 1 to form the second half of the hitch.

Emergency Halter

If you need to catch or lead a loose horse and don't have a halter, you can create one using a rope. An old lariat with a small loop at one end (called a honda) works nicely. However, a long flexible lead rope or even a piece of baling twine will do if you tie a loop in one end, using a nonslip knot like a bowline (see page 240), so the loop will retain its shape.

When catching the horse, slip the end of the rope over the horse's neck and grab it from the underside. Make a loop and push it through the honda at the end of the rope until it is large enough to slip over the horse's nose. This creates a makeshift halter, with the tail end of the rope serving as a lead. To turn the horse loose again, just slip off the nose loop, pull it back through the honda, and the rope will come off the horse's neck.

TRICKS *for*
THWARTING HORSES THAT UNTIE KNOTS

Here's how to prevent a horse from pulling on the free end of a quick-release knot and untying himself.

★ Run the free end down through the loop, to keep the knot from untying when pulled. Don't forget to slip the free end out of the loop when you're ready to untie it.

★ For extra insurance, use a rope that is long enough so the free end can go through the loop and clear down to the bottom of the post. Turn the knot around to the side and run the rope down the back of the post, then tie it again, around a bottom rail out of reach.

★ Another option with a long rope is to wrap it around the post where the horse is standing and then tie the end to the next post, so the horse can't reach the knot. This also keeps a horse from pulling away from the knot with enough force to make it difficult to undo.

★ Instead of tying the horse directly to a fence or other object, secure a loop of rope to a post at proper tying height, then snap the horse to the loop with a very short lead. There is no way the horse can undo this. Alternatively, use a 5- to 6-foot-long halter shank with a snap on each end: Just snap one end of the rope to the halter, run the rope around the post or through the tie ring (or whatever you are tying to), and snap the other end to the halter.

TYING & STAKING

The following tips give practical advice for easy and safe ways to tie or tether a horse.

★   ★   ★

Halters and Ropes

When tying a horse, always use a halter and rope that will hold securely and never break. A nylon halter (rope or webbing) is stronger than a cotton or leather one, and a three-ply web halter is stronger than one with thinner webbing in which the buckle holes may tear through. Any halter is only as strong as its hardware and stitching. If a buckle pulls out or breaks, or the snap on the lead rope breaks, the horse may suffer a serious accident, for example, falling over backward when he sets back. It's actually safer to securely tie the rope to the halter rather than snap it on.

THE RIGHT ROPE

When tying a horse, use a good-quality, strong rope that is not too small (³/₄-inch is usually a safe size). Make sure it's soft and flexible; a stiff or slippery rope may not hold a knot well, and a frayed rope may break.

Nylon rope is stronger and more weather resistant than cotton rope of the same diameter (a ¹/₂-inch nylon rope holds as much weight as a ³/₄-inch cotton rope). Nylon has a slicker surface, however, and is more apt to cause a rope burn if it slides through your hands or against the horse. The nicest kind of lead rope is a ⁵/₈- or ³/₄-inch soft cotton rope (or a mix of nylon and cotton fibers). Keeping it clean and dry will help prevent it from fraying or rotting.

Tying a Foal

When you halter-break a foal, don't tie him solidly at first. Stand at the post and hold him by the rope. Position someone behind him so he won't fall down or flip over backward if he pulls back; that person can push on his rump and encourage him to go forward again. As he learns to stand patiently, progress to looping the rope around the post and holding it, so you can still give a little slack if he pulls back and take it up again when he steps forward. He'll start to realize that he's being held by the post and must stand there, and that he'll slacken the rope if he steps forward.

The first few times you tie a foal, make it for just a few minutes and stay with him. Stand behind him so you can urge him forward if he pulls back. Gradually increase the length of the lessons as he gains patience and a longer attention span.

Tying with a Body Rope

When training a young horse (or any horse that may pull back on the rope), be careful to tie him so he won't injure his neck if he resists. Using a body rope is a good way to avoid injury and to teach the horse he can't pull free; it distributes the strain over his whole body rather than just on his head and neck.

Use a body rope to tie a young horse that's still learning.

★

Start by placing the rope around the horse's girth area and tying it under his belly with a nonslip knot like a bowline (see page 240), so it won't tighten up and pinch him when he pulls back. Run the free end between his front legs and through the halter ring; tie it, at head level or higher, to a sturdy fence post or something else secure enough to hold him. If you use a halter rope in addition to a body rope, tie the halter rope with slightly more slack, so that if the horse pulls back, most of the strain will be on the body rope.

Tying a Horse That Pulls

These methods will help thwart a horse from pulling back and hurting himself.

Use a neck cinch. If you have a horse that occasionally pulls back on the halter, outfit him with a cotton or nylon string cinch around his neck. Most of the pressure will be on his neck when he pulls back, and he won't be able to break the halter or hurt himself. Choose a cinch with rings on each

end, but no metal tongues in the rings (you can remove any tongues with bolt cutters). Place it around his neck just behind the halter, with the rings underneath. Instead of fastening the tie rope to the halter, thread it under the halter from his chin along his jawbones and out through the back of the halter's throatlatch. Tie the rope to the cinch rings under his neck with a bowline knot.

Tie high. When you tie higher than the head, the horse can't pull back as hard and is less apt to hurt himself. At this angle, he ends up picking his front end off the ground and soon learns that it's futile to pull back.

Cross-tie. Tie to two rings or posts (preferably higher than his head) on either side of him. By dividing the force of the pull in two directions, you're making it harder for him to pull back as strongly. For this method, use a well-padded halter and a neck collar (read about the neck cinch on the previous page). Run each side rope through the cheek ring of the halter (instead of the center ring under his jaw) and to the collar. Doing this will prevent the horse from pulling off the halter, which is a risk when you fasten a rope directly to the side of it.

Retraining a Habitual Halter Puller
Sometimes a horse develops the habit of flying back on the rope whenever he feels like it, perhaps because he managed to pull free earlier and thinks he can do it again, or because he had a bad experience with tying. Spoiled horses can be hard to retrain, but a method that often works is to reward good behavior and punish pulling back.

You'll need two people for these lessons. One person, the enforcer, stands quietly out of the way behind the horse with a straw broom or buggy whip. The other person stays near the front. When the horse stands quietly (with slack in the rope), the person in front praises and scratches him. If the horse pulls back, the enforcer gives a quick pop of the whip or a swat or poke of the broom on the horse's rump. When the horse steps forward, he's rewarded again. Most horses figure out that it's preferable to stand quietly and be rewarded than to be driven forward with a whip or broom.

Inner Tube Tie

A safe way to tie a horse that sometimes pulls back is to securely fasten a large tire inner tube to the fence, post, or wall, and then tie the horse to the inner tube. When he pulls back, the rubber will stretch without breaking, and it'll resume its shape afterward. This method is easier on a horse's neck than a solid tie.

Another way to use an inner tube (and a good one when tying young horses for the first time) is to attach strong ropes to two sturdy trees or stout posts at a height slightly above the horse's head. Tie the loose ends to opposite sides of an inner tube. This tight "picket line" with the tube in the middle has some give, and a horse can be tied anywhere along the line or to the inner tube itself. When he sets back, he can't pull hard enough to hurt his head or neck because of the height and flexibility of the line.

Tying to a stretchy inner tube prevents the horse from pulling hard enough to hurt himself.

Tying to a Tree

When tying to a tree, choose a strong, live one that can't be pulled over. Make sure a tied rope won't slip down its trunk. If possible, tie to a stout overhead branch. The horse won't be able to rub on or chew on the trunk, and he'll be less apt to hurt himself or pull loose.

When tying to a tree in the backcountry, protect the trunk from rope burns when you plan to leave the horse tied for a while. You can purchase a tree-saver strap, a broad band that fastens around the tree and is fitted with a loop that you tie the horse to. You can make your own version from an old car seat belt. First slip a solid metal ring over the seat belt, or braid a loop of rope around it, to give you something to tie to. Sew the ends of the belt together (or have your saddle shop or shoe repair shop do it), to create an adjustable circle when you buckle the belt. The wide nylon web will distribute the pressure if the horse pulls on it, preventing damage to the tree. Another simple trick is to use folded burlap sacks between the tree and the rope. The cushion of the sacks will keep the rope from burning or cutting into the bark.

Staking a Horse

A horse can be tethered to graze with a long rope fastened to a stake driven into the ground. Any kind of stake with a loop or hole in it will work, as long as it's not sharp on top. It should also stick up high enough for the horse to see it, so he doesn't inadvertently step, lie, or roll on it. Attach a swivel snap to each end of the rope, for fastening to the stake and the horse.

Don't tether a horse to a tree, or he'll end up wrapping the rope around it. The object you tie to must be very close to the ground. For this reason, it helps to teach him to be restrained by his foot, using a padded strap around a front pastern. This is much safer than staking him by the halter. In the latter situation, the rope comes up off the ground each time he raises his head, making him more prone to entangling himself. But when he's staked by a front foot, the rope stays on the ground and isn't as likely to wrap itself around a hind leg.

Preventing Rope Burns When Staking a Horse

To prevent rope burns, give your horse a few lessons before staking him. First accustom him to the feel of the rope by gently running it around his legs, over his back, and so on. If you plan to stake a horse by the head (which is not as safe as staking him by a front foot), also make sure he knows what it feels like when he steps on the rope and finds his head temporarily held down. Some horses panic and rush backward when they try to raise their heads and find they can't. Before staking a horse by the head, let him graze in supervised conditions a few times while dragging a lead rope attached to the halter, so he'll learn not to panic when he steps on the rope. Eventually he'll figure out that it's no big deal, and that by moving the proper foot he'll free his head.

A safe way to stake a horse and eliminate the risk of rope burn is to thread the rope through an old, large-diameter garden hose. This makes the tether stiffer, less likely to wrap around a leg, and less abrasive against the skin. To thread the rope, first pass a rigid wire through the garden hose, then hook the wire to the end of the rope and pull the rope through.

WATER TUB FOR A HORSE THAT'S STAKED OUT

Keep a horse's stake-out rope from spilling his water by setting the bucket or tub inside an old tire rim. To secure the rim in place, drive two stakes into the ground just inside the rim and across from each other. You can tie the rope to either stake, or to the rim.

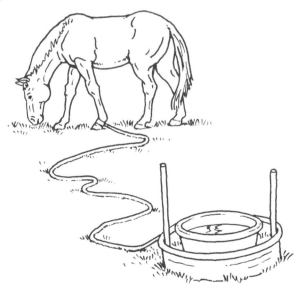

Staking a Horse When There's Nothing to Tie To

When you're riding or camping in open country and want to let your horse graze awhile but have nothing to use as a stake, you can tie a bulky knot in the end of your stake rope and bury the knot in the ground. Dig a hole 6 to 8 inches deep (use a stick or a combination pocketknife–screwdriver if you have no other digging tool), making the hole just large enough in diameter to stuff the knotted end of the rope in the bottom. Backfill the hole with dirt, tamping it firmly with your boot heel. The horse won't be able to yank it out if he's on the end of a long rope, because he'll always be pulling to the side instead of straight up.

Letting a Horse Graze While You Hold Him

If you want to let a horse graze for a while with his halter on but don't want to turn him loose or stake him, here's how to hold on to him and keep him from stepping on his halter rope. Make a loop (with another rope or baling twine) around the front end of his body to hold the rope up so it doesn't drag on the ground. Starting at his withers, run the rope or twine down around the front of his chest and then between his front legs and back up to his withers on the other side. Fasten it to itself to create a loose loop and then tie the ends to the mane at the withers. This keeps the loop in place so it won't slide forward over the horse's neck; and having it run in front of his chest keeps it from sliding back over his body. Thread your lead rope through the loop to hold it away from the front legs.

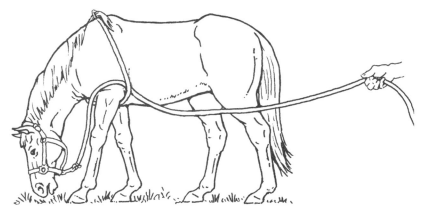

TIPS *for*
TYING SAFELY

Safety is the most important concern in tying a horse.

★ Make sure the footing where the horse is tied is good, so he won't slip and injure himself.

★ Tie the rope level with the horse's head or higher to make it harder for him to pull back with enough force to hurt himself by causing muscle strain, tendon or ligament damage in his neck, or even a dislocated vertebrae.

★ Never tie to a rail nailed to the near side of the post, because the horse may pull it off if he sets back. Another option is to tie to the post between the top two rails. Don't tie to wire fence or netting because the horse may catch his foot if he paws.

★ Unless you use a nonslip knot (like a bowline), don't tie a horse with just a rope around his neck. If a knot slips and the horse pulls back, he can strangle himself as the rope tightens. Using a halter is safer, because it distributes the pressure.

★ Tie short: 18 to 24 inches is plenty long to give a horse enough freedom of head movement so he doesn't feel claustrophobic and can step forward or back comfortably, and so he can't get a foot over the rope, or his neck under it.

★ Never tie with bridle reins. Sooner or later you'll have broken tack or a horse with an injured mouth.

ROPE BRIDLES

In an emergency, you might need to catch and ride a loose horse when you don't have a bridle. Here are a few options.

★　　★　　★

Makeshift Hackamore

If you must catch and ride a horse and have nothing but a rope (at least 10 feet long works best), you can create a hackamore with reins.

1. Fold the rope so both ends are together, creating two parallel lines. Tie a loose overhand knot (see page 238), forming a large loop where the rope is bent.

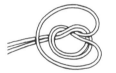

2. Fold the loop back around to encircle the knot.

3. Create a loose bowline (there'll be a double loop because of the doubled rope).

4. Adjust the two loops before you tighten the knot, so the longer loop will fit over the horse's poll like a headstall and the smaller loop will slip over his muzzle to create a noseband, leaving the two ends of the rope coming out at the back for reins.

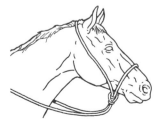

5. The knot under the jaw won't slip, so you will have a fair amount of control, especially if the nose loop is positioned low enough (below the bony bridge of the nose) to apply pressure on the soft tissues when you pull on the rope reins.

Makeshift Bridle

You can make a simple bridle (for use only in an emergency) just by looping a thin rope or baling twine around the poll, then crisscrossing the ends through the horse's mouth and bringing them back out as reins. When applying pressure on the reins, do it gently because there's no bit to distribute it evenly; the rope or twine can cut into the horse's mouth and tongue if you pull harshly.

Securing a Halter Rope without a Snap

Tying a lead rope directly to the halter is always safer than using a snap that could break. Many horsemen leave a rope permanently attached, sometimes even braiding it back onto itself a few inches, so that it will be permanent and unbreakable.

Another way to attach a rope that you can take off when necessary is to tie a loop in the end, using a nonslip knot so it will always stay the same size. Slip this loop through the halter ring from the back, then feed the working end of the rope through the loop from the back and pull the entire lead through. To remove the rope, merely pull it back through its own loop and the halter ring.

To attach this type of lead rope to a one-piece rope halter, slip the loop over the double loops on the halter (the ones you'd ordinarily snap a lead to), then pull the working end of the lead through the double loops.

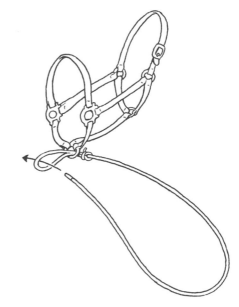

CHAPTER TEN

Handling & Riding

A well-mannered horse is a pleasure to work with. By helping your horse learn how to respond to different situations appropriately and confidently, you'll make the various aspects of your handling and riding experiences go smoothly.

CATCHING & LEADING

Being willingly caught and leading freely are very important aspects of every horse's training. Here are tips for dealing with a horse that needs more work in these areas.

★　　★　　★

Catching a Reluctant Horse

If a weanling or yearling is really hard to catch, keep him in a small, safe pen and leave a close-fitting halter on him for a few days until he's used to being caught. Using a breakaway halter is a good idea while you're training him: It's strong enough to restrain him when you take hold of it, but won't hurt him if he catches it on something. You can put a stronger halter on over it for leading and tying.

To train a horse that's very timid or stubborn and won't let you near enough to grab his halter, let him drag a long rope for a few days (again, do this only in a small, safe pen). Quietly catch him several times a day, moving slowly, in a nonconfrontational manner, until you can get close enough to grab the trailing rope. Eventually, the horse will realize he can't avoid being caught, and you'll be able to dispense with the halter and trailing rope.

TEACHING VOICE COMMANDS

It's easy to teach a horse voice commands if you use the same tone of voice for a specific cue. For instance, *walk* should be said in a gentle, soothing voice, to persuade the horse to relax and slow down from a faster gait. *Trot* should be spoken more sharply and crisply as an encouragement to move faster. *Whoa* should always be said strongly (but not shouted) and then reinforced by having the horse stand still for a moment so he learns that it means to stop and stay stopped.

Stopping a Crowder

If a horse does not respect your personal space and is always bumping into or shoving you as you lead him, use duct tape to secure a small dog brush rake (with blunt metal teeth) to the elbow of an old jacket. Wear the jacket each time you lead the horse, until he realizes that running into you is not so pleasant. Switch the rake to the opposite elbow and repeat the lesson from the other side, if necessary.

Moving Forward

For a yearling or older horse that hangs back or refuses to walk briskly beside you, use a stock whip or long switch as an extension of your arm to reach back and gently tap his hindquarters. This is more effective than pulling at his head, because a horse will instinctively pull back, resulting in a tug of war. For advice on leading a foal, see page 198.

Leading at the Trot

When a horse leads well at a walk, teach him to lead at a trot by moving your feet faster, trotting in place for a few seconds as you give the voice command to trot. With a lazy horse that needs more encouragement, you may have to reach back and tap his hindquarters with a stock whip. After a few lessons, he should start trotting whenever you give him the initial cue.

AIDS FOR RESTRAINT & CONTROL

There are many times you need the horse to stand still and behave. Here are some restraint methods that will help you control him.

★   ★   ★

Using a Chain Shank

Some rambunctious horses need more restraint than just a halter when you need to lead or hold them still for the veterinarian or farrier. A lead shank with a chain that can be passed through the side rings on the halter, over the nose, and hooked back to itself will often give the needed leverage to make the horse behave and stand still, or prevent him from dragging you faster than you want him to go. Still, it should not be over-used as punishment or cause pain, or else the horse may become more unruly. A relatively heavy, broad chain is more humane than a thin, lightweight one, which can cut into the skin.

A common mistake when using a chain shank is to apply continuous pressure. It tends to numb the tissue and, after a while, the horse will ignore the pull. Apply pressure only when necessary to remind the horse to stand still. That way, he'll figure out that by behaving, he'll be spared the chain's pinch on the bridge of his nose. Be sure to gauge the amount of pressure needed for each horse. Some horses are more sensitive than others, and harsh use of a chain will distress them and defeat your purpose.

Leading with a Chain

Some horses root at the halter or try to bolt, habits that often result from improper leading methods. A led horse should always have a little slack so that he can walk beside you in a relaxed manner. Remember, it takes two to have a pulling contest. Pressure on the halter to slow him down should be intermittent, not continuous. A few well-timed tugs are more effective than a hard, steady pull. Your right hand should grasp the end of the chain where it meets the lead line as you walk just ahead of the horse's shoulder, at roughly an arm's length from him.

With an aggressive horse that won't heed your command to slow or halt, use a chain over the nose that's adjusted to work the same way a choke collar works on a dog. It should not exert any pressure unless the horse starts going too fast or you ask him to stop. When he slows or halts, the chain will automatically release and loosen.

Using a Lip Chain

Another way to control a horse and make him stand still for medical treatment or some other procedure he dislikes is with a lip chain. Pass the chain on the end of a lead shank through the side rings of the halter, but instead of going over the nose slip the chain under the upper lip against the gum. When you apply pressure, the chain will press on certain nerves that block the release of adrenaline, thus creating a calming effect. Be careful: Used roughly, a lip chain can cut into the gum or lip, making the horse react adversely.

Applying a Traditional Twitch

When you need more than a chain shank to make a horse stand still, a twitch can come in handy. The traditional type has a wood handle, 15 to 30 inches long, with a loop of rope, leather thong, or chain attached to one end.

Be careful when you apply a twitch. A horse may strike out with a front foot or otherwise try to avoid it. Stand to one side so that if he rears, strikes, or slings his head,

CALMING A NERVOUS HORSE

To settle down a nervous horse while you hold him for the vet or farrier, gently but firmly rub his neck or the front of his withers with a circular motion. These are the spots where mares nuzzle their foals and where horses rub each other when grooming.

you'll be out of the way. To put it on a horse, reach through the loop and take hold of his upper lip, then slide the loop over your hand and around the lip. Next twist the handle until the loop tightens around the lip just enough to "tranquilize" the horse. Applying pressure on a horse's upper lip stimulates the release of endorphins, the morphinelike substances that act as the body's natural painkiller.

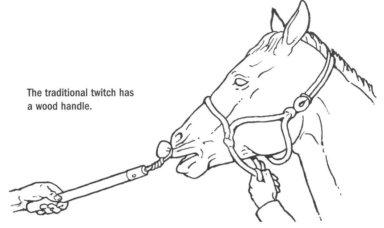

The traditional twitch has a wood handle.

If the horse starts to move or react to what else is being done to him, tighten the twitch a little more. Keep in mind that most horses will tolerate a twitch for a few moments, but not an extended time, so be ready to perform the treatment immediately after applying it. To remove the twitch, place your hand on the horse's upper lip and massage it with a firm, caressing motion as you untwist the loop. Continue rubbing the area until the horse relaxes, leaving him with a good attitude.

Using Other Types of Twitches

Use these judiciously, just as you would the traditional twitch.

Metal clamp twitch. This twitch is designed to clamp directly onto the upper lip and to the halter, leaving you with both hands free to work on the horse. Still, it's safer to have a helper hold it, because any kind of one-man twitch can be dangerous if it comes off the horse's lip. No longer under control, the horse may toss his head and turn the twitch dangling from his halter into a flying missile.

Safe one-man twitch. Make this simple twitch by tying a 15-inch-long piece of clothesline into a loop and clipping a double-ended snap onto it. Put the loop on the horse's upper lip (just as you would the loop of a traditional twitch), twist it tight, then clip the other end of the snap to a halter ring. This eliminates the need for an extra person to hold the twitch, and if it comes off there's no wood handle or metal tool to hit you or the horse.

Clothesline loop with double-ended snap

Twitch on horse

Hand twitch. When dealing with a quick, temporary discomfort (such as an injection or an application of wound medication) gripping the nose by hand is often adequate to keep the horse from moving. And it is generally easier than trying to put on a twitch when the horse is evasive. Grasp the nose and upper lip with your hand and twist or squeeze it. This restraint is very humane, because you can't apply enough pressure on his nose with your hand to hurt him. One disadvantage is that he can pull away, but once you grab hold of the upper lip, a horse will usually stand still.

Grasp the nose and upper lip.

USING AN EAR HOLD

This calming restraint, called "earring down a horse," can be either humane or inhumane, depending on how it's done. To do it properly, just cup your hand around the base of the ear, keeping your elbow bent so the horse won't pull your shoulder if he suddenly raises his head. Position your hand a fraction of an inch back from the edge of the ear, lining up your fingers and resting them on a ridge of cartilage at the top of the ear. Gently squeeze with your thumb to apply pressure on the ear cartilage, using a mild twisting motion to bend the ear's edge inward toward your palm. This hold can be very effective for short-term restraint and won't make the horse ear-shy. Be aware, though, that some horses may react strongly, striking out or slinging the head.

Skin twitch. Also known as a shoulder twitch, this restraint simply involves grasping a large amount of loose skin on the neck, just ahead of the horse's shoulder. Squeeze the skin as hard as you can, twist it a little, and roll your knuckles forward to pull a fold of skin over your fingers. Holding a horse this way tends to immobilize him, keeping him from moving or striking out.

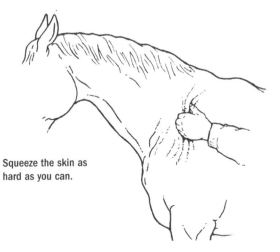

Squeeze the skin as hard as you can.

Using a Stableizer

This tool consists of a looped cord that goes over the top of the head and under the top lip, next to the gum. A tiny set of pulleys allows you to tighten the loop by pulling on the cord handle. When tightened, the device activates pressure points behind the ears and beneath the upper lip, blocking out pain so the horse feels calm and relaxed. The portion

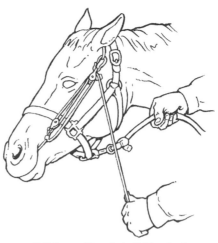

Pull the cord handle to tighten the loop.

of the cord that goes under the lip is covered with plastic tubing so it never cuts into the gum, making it more humane than a lip chain.

With a Stableizer, it's important to make sure the horse is fully relaxed before you begin working on him or giving him medication. This usually takes about two minutes from the time you fully tighten the loop. The clues are obvious: The horse's head lowers, his ears flop, his eyes look sleepy, and his lower lip droops.

Wait for the horse to fully relax before working on him.

USING A BLINDFOLD

Horses can sometimes be temporarily immobilized with a blindfold, and it's worth a try when dealing with a horse that won't tolerate a twitch or other restraints. A blindfold is also useful for leading a horse out of a burning barn, given that the horse's usual reaction is to panic and run back to his stall.

A traditional way to blindfold a horse is to stuff a jacket or grain sack under his halter, covering his face. Although this type of blindfold works (and it may be the only thing you have in an emergency), a large flapping object can alarm the horse unless you put it on very carefully so that it doesn't slip.

You can make a quick and less spooky blindfold from four clothespins and a towel folded in half to fit the area you need to cover. Fasten the towel to one side of the halter with two clothespins, then spread it across his face in one easy movement and clip it to the other side with the other two clothespins. The blindfold should hold very well unless the horse rubs the pins against something. When you're ready to remove it, unclip all four clothespins and lift the towel, again in one easy motion to avoid startling the horse.

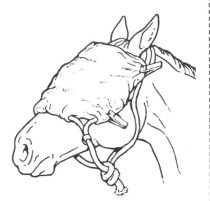

Blindfold clipped in place

BRIDLING A HORSE

Being bridled is an exercise that need not be uncomfortable or traumatic for a horse. The key is to behave carefully and confidently. Here a few effective bridling techniques and practices.

Opening a Horse's Mouth

If a horse is hard to bridle because he keeps his teeth clamped shut, put your finger into the corner of his mouth to encourage him to open up for the bit. If he still refuses, press on his gums, in the space between the molars and incisors.

SWEETENING THE BIT

For a horse that's especially stubborn about taking the bit (throwing his head up in the air to avoid it, for instance), offer him a reward rather than trying to fight with him. Dab molasses on the mouthpiece and let him lick it off. Even if it takes several sessions, wait for him to grow eager and willing before you actually try to bridle him again. You'll have an easier time slipping the bit into his mouth once he realizes it tastes good.

Fitting the Bit

Adjust the headstall so the bit hangs properly in the mouth; not too tight that it presses hard on the mouth corners, and not hanging loose enough to bang his teeth or let him get his tongue over it. A curb bit should fit at the corners of the mouth with no wrinkles, whereas a snaffle should create one small wrinkle.

Choosing a Curb Bit

A curb is generally more severe than a snaffle, because it works like a lever instead of just pulling on the corners of the mouth and gums as the snaffle does. Pulling the reins attached to the shanks of a curb tips the bit in the mouth, which affects the horse in several ways: It tightens the curb strap or chain against the chin; it pulls down on the headstall, applying pressure on the top of the head; and it pushes against the roof of the mouth (if the port in the mouthpiece is very high).

Some curb bits are more severe than others. A curb with a low, wide port is mild because it allows the horse a lot of tongue room and doesn't touch the roof of the mouth. The length of the shanks also affects the severity of the bit. The proportional length (the part above the mouthpiece compared to the part below) determines the leverage on the jaw. If the upper portion is only slightly shorter than the lower, the bit is quite mild, exerting less leverage. But if the upper part is quite short and the lower relatively long, a lot of leverage will result from even a slight pull on the reins. When using this type of bit, never pull hard on or jerk the reins.

Selecting a Snaffle Bit

Make sure the snaffle fits the horse's mouth properly. If the mouthpiece of a straight snaffle is too long, it'll slide around in the mouth too much. A broken snaffle that's too wide for the mouth will buckle and poke the roof of the mouth when the reins are pulled. If a horse starts evading the bit action of a broken snaffle, it may help to switch to a French snaffle with a three-part (or two-jointed) mouthpiece.

Removing a Bridle

Make sure your horse lowers his head before you try to take off the bridle. If he raises it, the bit will catch on his lower front teeth and cause him pain. This unpleasant experience can make future unbridlings more difficult, because his instinct will be to throw his head up in the air to avoid the pain he's anticipating. To eliminate the problem, teach your horse to lower his head on cue by rubbing his forehead or applying a gentle downward pressure on his lead rope if he's wearing a halter under the bridle.

POINTERS *on*
TRAINING WITH ELASTIC REINS

When you're first getting a horse used to wearing a bit, try using elastic reins hooked to a surcingle or to your saddle. When he roots at the bit, it won't hurt his mouth and the reins will gently pull his head back into proper position. You can make elastic reins from the inner tubing of an old car tire (cut into strips) or bicycle tire. Fasten a snap to the bit end of the rubber strip by first cutting a slit in the rubber, then feeding the end of the strip through the loop end of the snap and running it back through the slit. Pull tight. Next fold over the girth end to create a loop large enough to put your girth or surcingle through. Again, slit the rubber strip, this time to run the snap end of the strip through. The length between the bit and girth should let the horse hold his head in a relaxed but semicollected position.

To make adjustable reins to fit any horse you may be school- ing, use a shorter piece of rubber and attach it to an old belt or other piece of leather with a buckle. Rivet a metal ring to the end of the leather, and attach the rubber inner tubing to the ring, using a slit to put it through itself, as you did when you attached the snap to the other end. Place the leather end around the cinch ring or girth, buckling it at the proper length.

MOUNTING

Here are a few methods that make it easier for a rider to mount a horse bareback as well as with a saddle.

★　　★　　★

Mounting Bareback

The traditional and proper way to mount bareback (without a mounting block) is to gather the reins as you stand at the horse's left shoulder, place your hands on the horse's withers and spring upward, using the strength in your arms and the push from your legs to maneuver as much of your upper body as possible across the horse's back right behind his withers. Balance your weight on your abdomen. If the horse starts to move at this point, check him with the reins. Then pivot around, swinging your right leg over the horse, and move into the mounted position.

A short rider with a tall horse may find it impossible to mount in this fashion. In this case, grasp the mane at the withers for a handhold and swing your right leg up over the horse's backbone. With strong arms and a bit of momentum, even a short rider can mount a tall horse this way.

Using a Mounting Block

A very simple way to ease the challenge of mounting is to stand on a bale of hay or straw. Teach your horse to stand calmly beside the bale while you mount. If you've taught him the meaning of *whoa* and insist that he stand still until you have mounted and given the signal to move, a mounting block makes a very safe aid.

WARMING UP

Always warm up the horse with slow work before asking for speed.

★   ★   ★

Stretching and Flexing

To help a horse perform his best, with less risk of injury or stiffness following a ride, warm him up properly before every strenuous ride or exercise. Begin at a brisk walk, asking the horse to cover as much ground as possible without breaking into a trot. Along with warming his muscles and stimulating natural lubrication of the joints, the exercise increases blood flow to the tendons, making them more elastic.

After several minutes of walking, progress to a slow trot or jog for a moment, then go back to a walk. Repeat this routine a few times. If the horse is fit, his warm-up period can include changes of direction, some circles, and short periods of alternating a working trot and a walk and then a canter and a walk. Any greater speed early in your ride will defeat the purpose of the warm-up.

Knowing When He's Warmed Up

If the horse is just starting to break into a light sweat (his coat feels slightly moist to the touch), your warm-up has been about right. If he breaks into a real sweat (or a lather), it's been too much. Drop back to slower exercises in subsequent warm-ups until he becomes more fit.

HANDLING A HORSE FROM BOTH SIDES

People have traditionally handled horses from the left, or near, side because most of us are right-handed and hold the lead rope in the right hand. But when you always approach a horse from the near side, you inadvertently train him to be uneasy when you're on his right, or off, side. Work with a horse from both sides so that he'll be at ease in either case. In the mountains, for instance, it's easier to mount from the uphill side, whichever that happens to be.

Don't Feed Grain Right Before Riding

A horse may develop digestive problems when exercised too soon after a large feed of grain. Blood that's needed for digestion is suddenly routed to the working muscles instead of the stomach. Sugars and starches in the grain that should be digested and absorbed in the small intestine may enter the large intestine undigested. Fermentation of these nutrients will alter the microbe population in the large intestine and may lead to colic or founder.

Grass hay can be safely fed right up until the horse exerts himself, but grain should be given several hours before or after work. Large feedings are best given in the late evening, after the horse is through working and has cooled out, and when he'll have more time for proper digestion.

GOING THROUGH THE PACES

The more in tune you are with the way a horse moves, the easier it will be to communicate and maintain the different paces you want to set when you ride.

★　　★　　★

Speeding up the Walk

The horse's legs move in repeated sequence: right front, left hind, left front, right hind, right front, and so on. Because most of his propelling power comes from his hind legs, you can encourage him to push off harder, thus lengthening his stride and walking faster, by stimulating each hind leg at the moment it's pushing off the ground. The simplest way to do this is to watch the horse's shoulders. Each front leg comes forward a split second before the opposite hind; as the left shoulder moves forward, his left front leg is in the air and ready to land, and vice versa. It happens so quickly that you'll be able to achieve the proper timing if you squeeze each side, exerting the pressure through your upper calf muscles, just as the opposite shoulder moves forward. This method works better than simultaneously squeezing both legs, because once a hind leg is in the air, there is little the horse can do in response to your squeeze.

Eventually you'll feel the rhythm with your legs and body, and you'll be able to squeeze automatically without watching the shoulders. At that point, all you have to do is swing with the horse, flexing at your hips

GETTING IN THE MOOD

A 5- to 10-minute warm-up not only improves circulation by increasing oxygen intake and absorption, but it also prepares the horse mentally for the task ahead. He'll be more apt to do his best and less likely to try foolish, goof-off stunts that could increase the risk of injuring "cold" muscles and tendons.

GIVE A HORSE A BREAK FROM TRAINING AND CONDITIONING

Horses often learn and perform better when they have short vacations now and then. Often you'll find that after a day or two off, a horse will be sharper in his responses and able to perform a new movement even better than when you left off. The best training program combines daily sessions with a few well-timed breaks to allow the horse time to digest what he's learned and come back to his lessons refreshed.

and waist and rolling your weight in the saddle a little as you alternately squeeze your legs. In this way, you'll be pushing the horse with your whole body (especially your legs and seat bones) as you move in rhythm with him.

At the same time, keep light contact with the bit so you can stop the horse from breaking into a jog. You can usually feel or sense when he's about to do this. Each time you check him, continue squeezing with your legs, so he'll know he should walk faster and not slower.

Sitting the Trot

The key to sitting a trot comfortably is proper body position and muscle relaxation. When you try to grip with your knees, you interrupt proper movement and don't stay in sync with the horse. If you grip with your thighs, you can't sit deep in the saddle. And if you grip with your calves, the horse may take it as a signal to increase speed. You want relaxed legs that dangle alongside the horse.

To gain the proper position, trot slowly without stirrups. Sit heavily in the saddle with your weight on your seat bones and allow your legs to hang naturally, straight from the hips, toes angled downward. Lean back slightly and relax, but don't sit too far back, or you'll feel more thrust from the hind legs and bounce around more. Now let your hips and lower back gyrate and move freely up and down with the horse's motion (a relaxed, rolling movement). You need a lot of motion in your hips and lower back to follow the horse's movement while never losing contact with the seat of the saddle. If you practice first at a very slow trot, you'll soon have the hang of it.

When you can keep a good seat with your legs hanging, put your feet back in the stirrups. If your position is correct (sitting in the middle of the saddle), you should be able to find the stirrups just by raising your toes. Rest on the balls of your feet without putting any weight on the stirrups or pushing your heels down. That would stiffen your legs, causing you to lose contact with the saddle and start bouncing. If you relax your calf muscles, they'll stretch, and gravity will pull your heels down to the proper position.

RIDING RING GATE

For a simple riding ring gate, hang a lightweight pole across the opening of the ring at a height a rider can reach to open without dismounting. Large rings or even old stirrup irons fastened to the sides of the posts can serve as pole holders.

Striking the Right Pace

When a horse is allowed to select or change his gait on his own when ridden, he'll usually make the transition at the point it would take more energy to continue in the present gait than to use a different one. For instance, he'll start to canter when it takes more energy to extend his trot (in order to go faster). Likewise, he'll drop to a trot rather than spend energy to collect (slow down) his canter.

Terrain also influences a horse's choice of gait. On uneven surfaces, the trot is more stable than the gallop because it supports the horse at every step (with a leg on each side and at each end of the body on the ground), whereas in a gallop or canter only one foot is on the ground for part of each stride. If you do a lot of cross-country riding, get to know your horse's comfort range for each gait, and allow him to "shift gears" accordingly.

CAREFUL COOLING OUT

To avoid stiff muscles or more serious problems, cool the horse out after every hard ride.

★   ★   ★

Stopping Gradually

Cooling out is very important following any strenuous exercise. Human athletes know better than to stop abruptly after exertion. That's why runners continue after crossing the finish line, slowly drifting down to a walk, flexing their muscles and squeezing out the waste products, and flushing out and using up most of the accumulated lactic acid.

A proper cooling out not only does the same for horses, but it also dissipates accumulated body heat, returning a horse's core body temperature to normal. As long as a horse is moving, the evaporation of sweat helps cool him. Depending on the length and degree of the workout, a horse will need a cooling-out period of 10 to 30 minutes (or longer after a race, a long endurance ride, or several hours of speed work in an arena). Progressively slow the work, easing the horse's body systems back to normal. After a long ride, walking the last 30 minutes should be sufficient. If you were galloping, do some cantering, then trot for 3 or 4 minutes, and finally, walk the horse until he stops sweating and dries off.

WHEN A HORSE BREAKS INTO A SECOND SWEAT

A horse that's not properly cooled out after prolonged exertion may dry off and seem cool (although a thermometer will tell you otherwise), but will break out in sweaty patches over the shoulders, neck, armpits, and flanks again after you've put him away. Always check him a half hour to an hour after a cooling out. If he's still overheated, lead him around some more until he stops sweating and is completely dry, and his temperature is back to normal. When in doubt, take his temperature.

When you can't keep riding the horse to finish cooling him out, dismount and loosen the cinch or girth, and lead him briskly for a while until he's dry. Unsaddle him and lead him some more, slowly reducing the speed of his walk over the next 10 to 15 minutes. Finally, groom him and massage his muscles.

Cooling Out at the Trot

The traditional way of cooling out a horse (walking after a workout) is not as beneficial as trotting and then walking. After hard work, the body returns to normal more rapidly when the slowdown is gradual and the blood keeps pumping and clearing away waste products. If he's been working at racing speed (or chasing cows or doing any steady strenuous work), drop him to a gentle canter, then to a trot. After trotting several minutes, bring him to a walk. The longer and harder the workout, the more time it will take to cool him, and much of this can be done at a trot.

Why Cool Out?

A proper cooling out, which enables all the horse's body systems to gradually return to normal, accomplishes the following.

★ Flushes waste products from working muscles

★ Uses up any remaining lactic acid (so it won't be stored in the muscles and cause soreness and stiffness)

★ Clears out any breakdown by-products of muscle tissue exertion

★ Allows the heart to gradually return to its resting rate as the volume of blood passing through the muscles diminishes

★ Keeps fluid from accumulating in the legs and joints and causing swelling

WINTER RIDING

Here are several tips to make riding in cold and snowy climates easier, safer, and more enjoyable.

★ ★ ★

Snow-Free Feet

To keep snow from balling up in the horse's feet, make the surface nonstick by greasing the bottoms of the hooves with lard, shortening, butter, margarine, cooking spray, mineral oil, petroleum jelly, or (for the longest-lasting coating) ski wax.

Saddling Up

You must approach saddling a little differently in the winter.

★ Always clean the horse's back before saddling, even if you don't have time to remove all the mud from the rest of his long coat. Dirt under the saddle pad will make his back sore.

★ You may have to make adjustments in tack. A thick winter coat or increased body weight may make a difference in the number of holes you'll take up on the cinch or girth.

★ When your horse has long hair, be careful to not catch it in the cinch ring, latigo, or girth buckles, an experience that may make him resent being saddled. A good precaution is to slip a hand between the horse and the cinch or girth as you tighten it.

Warming the Bit

In cold weather keep bridles in the house, as an ice-cold bit placed in his mouth is miserable for the horse. If a bit cools down too much before bridling, warm it in your hands. To keep a bit warm for a while even in very cold weather, try this method: Wrap it in a washcloth that you immersed in very hot water and wrung out, then place it in a plastic bag to keep the moisture in, and wrap a hand towel around the whole thing.

TIPS *for*

WINTER RIDING

Be especially careful when riding on snow or ice.

★ When traveling on slippery terrain or in deep snow, keep the horse collected so he'll have better balance and agility and be less likely to fall or flounder. Still, don't overly restrict the use of his head and neck. He must be able to move them freely to balance himself on difficult footing. Be mindful that frozen ground can also be treacherous because the horse's feet can't dig into it.

★ When descending a hill, head straight down it, never sideways. A horse can slip and slide all the way to the bottom and still keep his feet under him if he's going straight. But sideways his feet may slip out from under him and cause him to take a bad fall.

★ If your horse is reasonably sure-footed, it's usually safer to stay mounted than to get off and fall down in front of him or have him slide into you. If you must dismount, stay well out of his way. Where the footing is bad, move off the icy trail and into undisturbed snow, which will be less slippery.

Dressing for the Cold

In cold weather keep windchill in mind, and dress appropriately in clothing that's both warm and lightweight so you can move freely. Remember, when you and your horse are moving, the effect on your exposed skin is the same as when the wind blows. While traveling at a brisk trot, for example, your body loses heat at the same rate it does in a 10-mph wind. In 15°F/–4°C weather, that's the same as being exposed to a temperature of –2°F/–19°C.

★ To keep your feet warm, wear riding boots that are loose enough for heavy socks, with wiggle room for your toes. Boots that are too tight will restrict circulation in your feet. Don't wear bulky footwear that could hang up in a stirrup.

★ To prevent cold air or snow from blowing up your jacket sleeves (if the cuffs are not elastic), and to protect the area between your gloves and sleeves, cut the tops off old tube socks and slide them over your wrists.

★ To keep your ears warm and protect the back of your neck, pull a stretchy acrylic stocking cap over your helmet. It's warmer than a wide knitted headband worn with a hunt cap, because the pocket of air inside the helmet serves as additional insulation.

★ Don't forget a scarf to protect your neck and face, and lightweight, flexible gloves that let you feel the reins.

CHAPTER ELEVEN

Miscellaneous Matters

There are countless valuable hints to be gleaned
from other horsemen, many more than can fit in this book.
Here are just a few assorted ideas that may be useful at
some point when you're working with horses.

FLIES & OTHER PESKY INSECTS

During summer there are many pests that irritate horses. Here are ways to make bug season less miserable.

★   ★   ★

Fringed Swishers

Before the days of insecticides and repellents, horsemen augmented the horse's own weapons, his mane and tail, by giving him extra swatting and swishing power. This still works today. You can tie long strips of fringed denim or heavy cloth to a halter, bridle, or harness, so the horse can shake his head or wiggle to brush the flies off his face and body more easily. Here's how to make a fringe-style swisher for your horse's face.

1. Cut off the leg of an old pair of jeans at the knee, or a little higher (the cut piece needs to be about 20 inches long).

2. Slit the leg from top to bottom to form a flat piece of denim. To create an opening for the horse's ears, make a slit across the top (the cut end, not the one with the thick hem) to within about an inch of each side.

3. On the end with the thick hem, fringe the cloth from the bottom to within an inch or so of the top slit. The extra weight of the hem makes this fly shaker more effective, by keeping the strips vertical after the horse shakes his head.

4. Insert the horse's ears through the top slit, and secure the swisher by tying the two end fringe strips around the throatlatch. The denim is sturdy enough to last, but not so sturdy that it'll strangle the horse if he catches it on something; it'll pull off over his ears or tear away at the throatlatch.

★

Fly Mask

Today you can purchase many styles of fly masks and bonnets at tack shops. The masks cover the eyes with mesh that allows the horse to see but keeps flies out. Some masks protect just the eyes, but others also shield the ears and cheeks. Some even have fleece-lined ear openings and edges to keep flies from crawling in under the mesh. Fly bonnets with sewn-in soft mesh ear covers offer the same protection.

When putting a fly mask on, start with it on the offside of the horse's head and slip the crownpiece over the off ear, as you would when bridling. Bring the mask over the face and place the crownpiece over the near ear, then fasten the throatlatch. (If the fastener is Velcro, first make sure the horse is used to the sound of it pulling apart, so he'll be prepared later when you take off the mask.) Be sure the mask fits comfortably over your horse's eyes and is snug, but not tight, around his face so that flies can't sneak inside.

Homemade Ear Net

You can use an old pair of panty hose to shield your horse's ears. Make a 4-inch slit from the top of the waistband downward, so it won't be so tight, and pull the panty hose over the ears and poll. Tie the legs under the throatlatch, or tuck them under your bridle if you're riding. The horse may look a little funny, but he won't be bothered by pesky little insects in his ears.

FEATHER FLY WHISK

To brush flies off your horse while riding, take along a feather duster. You'll be able to reach his ears, neck, and rump with this simple tool. If you wish, attach a loop of string to slip around your wrist so you won't drop the duster.

Belly Net

To protect a horse's underside from being chewed up, you can cover the belly area with mosquito netting. One way to use the netting is to fasten it to a regular sheet blanket with strips of Velcro. First measure your horse to be sure the netting will fit properly, then sew one part of the Velcro to the inside bottom of the blanket and the other part to the netting.

Another method is to make a tube-style fly sheet from mosquito netting (you'll probably need about 4 yards). A body tube like this works nicely on a mare, but not so well on a male horse. Measure your horse's back, underside, and girth, then cut a piece of netting to fit around the barrel area. Sew an elastic strip on both the girth and flank ends, then sew Velcro along the remaining two edges (they will come together along the horse's back). You can also stitch on an adjustable strap to buckle or snap in front of the chest to keep the tube from sliding back.

Fly-Resistant Horse Clothing

Some fly masks and light sheet blankets are permeated with a fly repellent that remains effective for two years or longer. The specially treated fabrics repel or kill most flies, mosquitoes, and other insects, as well as spiders and ticks. This type of fly control is a spin-off from the military use of permethrin-impregnated uniforms to protect soldiers from biting and stinging insects. Permethrin is an EPA-approved insecticide used in many sprays and wipes and on vegetable and fruit crops to prevent insect damage. It is highly toxic to insects and other arthropods, but one of the least toxic insecticides to mammals.

HINTS *for*
APPLYING FLY REPELLENT

★ Before spraying actual insecticide on your horse, spray him with plain water until he no longer fears the sound.

★ Always take great care never to get fly spray in a horse's eyes.

★ Don't forget to use fly spray or repellent on the lower legs. When flies land on a horse's legs and bite him there, the horse tends to stomp his feet to dislodge them, which can be hard on the hooves and legs.

★ For sensitive areas, such as the udder, sheath, and face, spray the solution onto a soft cloth or mitt to apply it, or use a repellent cream instead of a spray.

★ If you don't have a fly repellent cream to protect your horse's face, try filling a well-washed, empty roll-on deodorant bottle with spray solution. This will make it easy to apply solution around the ears and eyes without making a mess and getting it on your hands. And you'll be less apt to spook or upset a sensitive horse than if you use a cloth or mitt. Another option is to dip a feather in the repellent and dab gently around the eyes.

HANDY APPLICATOR MITTS

To make a mitt for spreading repellent over the horse's body, cut a 7-by-18-inch rectangle from an old terry cloth towel. Fold it in half with the shorter sides matched up, then sew the 9-inch sides closed, leaving one open end for inserting your hand. Another option is to cut the sleeve of an old sweatshirt to the appropriate length, turn it inside out, sew one end shut, and turn it right side out again. Whichever version you make, use a plastic bag inside the mitt to protect your hand from any fly repellent that soaks through the cloth.

Other Fly-Fighting Aids

Here are additional devices for protecting your horse.

★ Nylon fly-shaker strips can be attached to a halter or bridle, and some are made as a browband with a quick-release feature that breaks away if it catches on something. Some shaker strips are impregnated with fly repellent that lasts several months. Fly nets with longer nylon strips can be buckled onto a harness.

★ A fly sheet (horse blanket with a cool, open-weave mesh) covers the body and keeps flies from biting. Because it allows air flow and sweat evaporation, it won't make the horse too warm.

★ Fly boots are helpful for horses that do a lot of stomping to dislodge flies from their legs.

Controlling Flies around the Barn

One way to kill flies in the stable and barnyard is to lay out a wet gunny sack in a place where horses have no access to it, then pour a mixture of syrup and insecticide on it. Flies will be attracted to the sweet syrup and perish. When the sack is covered with dead flies, hose it off and reuse it for the same purpose.

HOOVES & SHOES

Choosing the best style shoe for your horse is one of the most important aspects of horse care. Some factors to consider before you decide are the time of year, your location, and how often you ride.

★　　★　　★

The Growing Season
The hoof wall grows faster in spring and summer than in winter, due to nutrients in green grass or seasonal differences in body metabolism. You will have to trim the hooves more often if the horse is wearing shoes, or if he's not exercising enough to keep his feet worn down while barefoot.

Promoting Faster Hoof Growth
To stimulate faster hoof growth, for example, to help a horse with severely broken and chipped feet and not much hoof wall left, some horsemen rub a mixture of equal portions hoof dressing and Reducine liniment into the coronary band once a day. The ingredients in this liniment work as a mild irritant, which encourages a hoof to grow.

Toe Tips for Winter
Snow and ice build up more readily in the bottom of a shod hoof than a bare foot, because a shoe changes the shape and angle of the bottom surface, hindering its natural self-cleaning action. Still, horsemen who ride in winter may prefer to have shoes on their horses. Also, horses out in large winter pastures, where they must paw through crusted snow to get to grass, may wear their front toes excessively when left barefoot.

Old-time horsemen often employed just the front half of a shoe, called a tip shoe. It can be a good compromise, providing some protection for the foot while eliminating the circular shape of a shoe, which tends to trap and hold packed snow. This type of shoe can be fitted with toe calks or spots of borium welded on each side of the toe to give good traction on ice or frozen ground. The calks or borium spots shouldn't protrude much, however, or they'll throw the foot off balance, as there can be no corresponding heel calks to keep the foot level.

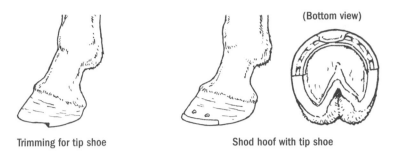

(Bottom view)

Trimming for tip shoe   Shod hoof with tip shoe

Almost any shoe will work as a tip shoe, but a thin one requires less special trimming of the foot (the toe must be trimmed more than the back half, to keep the foot level when the shoe is set). Even a used, partly worn-out shoe will do, especially if it has borium, which will keep it from wearing out more. To transform the used shoe into a toe tip, cut it with a hacksaw or a cutting torch (the latter is quicker), and then smooth the ends with a grinder.

Make sure the back portion of the hoof wall is smoothed and rounded so it won't split or chip, since it has no shoe to protect it. The place where the shoe ends should be trimmed to fit perfectly, so there's no unevenness and just a smooth ground surface.

Pulling a Shoe

If a shoe starts to work loose and the farrier can't come immediately, you may have to pull it yourself so it won't catch on something and injure the horse, or slip out of place and bruise his sole. It's fairly easy to remove without breaking the hoof wall if you have shoeing tools (a shoeing hammer, clinch cutter, rasp, and nippers), but if necessary you can use a carpenter's hammer, a flat-edged screwdriver, and pliers or vice grips.

First unclinch the nails. With a hammer, drive the chisel side of a clinch cutter, or a screwdriver, up between the hoof wall and each nail end to pry it up and straighten it. (Be careful not to cut into the hoof wall with a screwdriver.) Cut off the straightened nail ends with nippers. If you don't have the tools you need to unclinch the nails, just rasp off the ends.

If you are unable to cut or rasp the clinches, you can still remove the shoe if you have some kind of nippers, especially if the shoe is already loose. Hold the hoof in regular shoeing position: between your legs if it's a front foot; across your thigh if it's a hind. Starting at the heel, place the nippers between the shoe and the hoof and pull outward and downward to pry and loosen the shoe. Alternate down each branch of the shoe as you work toward the toe. It takes more strength and leverage to pull the clinches loose and on through the hoof wall than it does to cut them, but they will straighten out as you pull the shoe. To keep from breaking the hoof wall, go slowly and pull out each nail as you get it loose enough to grasp with nippers, a hammer claw, or pliers.

To pull a loosened nail, you may have to gently pound the shoe back down against the foot so the head will protrude enough to grab. If part of a broken nail remains in the hoof wall, pull it out with nippers, or it may snag the horse's other leg when he's traveling. Save a shoe that's not badly worn; your farrier can probably put it back on again.

Protecting an Unshod Foot

If it might be a few days before a shoe can be put back on, and the horse cannot be kept on soft footing, you'll need to protect the bare foot so it won't bruise and chip. An Easyboot or some other type of protective hoof boot will work, but if you don't have one, duct-tape a folded towel, diaper, or other padding to the bottom of the foot. Also tape around the edge of the hoof to keep it from splitting.

SHORT-TERM TRAVEL ON PAVEMENT

If you're only going for short trips on pavement, synthetic shoes or protective hoof boots may hold up well enough to provide the traction and cushioning your horse needs.

Riding on Pavement

Metal shoes can be dangerous on pavement. Besides compromising the horse's traction, they also create more concussion on the feet and legs and wear out quickly. Rubber shoes, which you can cut from old car tires with a band saw or a reciprocal saw, work much better.

THE LAUNCH OF RUBBER SHOES

During the 1976 west-to-east Bicentennial Wagon Train journey, teams of horses traveled mainly on paved roads, wearing out conventional shoes every 100 miles or so. Because the teams completed about 25 miles per day, the continual need for new shoes created a serious problem: Hooves won't hold up with that much nailing. Borium-studded shoes lasted as many as 300 miles, but the hooves still did not grow fast enough to be reshod every two weeks. Farriers accompanying the wagons kept experimenting and had their best luck with rubber shoes, which they cut from discarded car tires and nailed to the foot. The full rubber covering provided better cushioning and traction and lasted much longer.

However, one drawback to consider is that horseshoe nails can work out of the rubber or break off relatively quickly because there's more movement with a rubber shoe than a metal shoe. To keep nail heads from wearing off, recess them so they don't come in contact with the pavement. Just cut away the rubber where each nail will be positioned so the head will rest against the cord of the tire. You can place a small flat washer against the rubber under each nail head to reduce movement and keep the nail from pulling through. Another method is to flatten out the nail heads before nailing them. An all-rubber shoe should last long enough to protect and cushion the foot over many, many miles of pavement.

Rubber Pads for Metal Shoes

A rubber pad cut from a tire can be used in conjunction with a regular horseshoe to cushion better and provide more traction on hard surfaces. After the metal shoe is shaped to fit the hoof, use it as a pattern to cut out a rubber pad that's the same size and shape as the bottom of the foot. Trim the outside edge of the rubber (where the shoe will go) down to the tire cord. When the shoe is nailed on, the thinner cord portion under the shoe keeps the pad securely in place, while the thicker tire tread inside the shoe touches the ground, taking as much weight as the shoe and giving better traction on pavement.

STOPPING BAD HABITS

Sometimes a horse will develop an unwanted or even eccentric habit. Here are a few tips to thwart such behaviors.

★ ★ ★

Pawing and Kicking

When your horse paws or kicks in his stall (as some horses do when impatient at feeding time), a simple and safe way to mete out "punishment" from outside the stall is with a squirt gun. Squirting the horse the instant he misbehaves startles him enough to halt the undesirable activity. It also makes a big impression on him because the punishment seems to come from out of the blue. He'll think he caused it, rather than you, and that's much more effective than yelling at him or trying to spank him. It usually takes only one or two applications from the water pistol for most horses to halt the bad behavior.

To stop a horse that continually kicks at his stall wall, attach a short chain around a hind pastern with a soft, padded leather strap. (The short chain used as a rattle on a gaited horse's front feet, to make him lift them higher, can be used for this purpose.) The unfamiliar, sometimes startling sensation the chain creates when the horse strikes the wall is usually unpleasant enough to stop chronic kicking.

CHEWING TREE BARK

To deter a horse from eating the bark of trees in the paddock or pasture, cut open some burlap bags, then wrap and securely tie them around the trunks. Use as many bags as needed to cover the area of the trunk that your horse can reach.

TIPS *to*
PREVENT BLANKET CHEWING

Sometimes a horse will continually chew his blanket or bandages. Here are some things you can do to dissuade him.

★ Spray Bitter Apple (available through pet shops and pet-supply catalogs) on the item he's chewing. It has a very bad taste, but leaves no residue and won't stain blankets. You can spray it on a mare's mane or tail to stop her foal from chewing on it.

★ Dab mentholated rub on the parts of his blanket he can reach; most horses don't like the smell.

★ Tie a bunch of short pieces of baling twine to the lower part of a horse's halter, creating a thick twine "beard" that he won't be able to bite through to rip at his blanket. This device is cheaper and often more effective than a commercial bib.

★ Mix powdered laundry soap with just enough water to make a paste and apply it to the outside surface of your horse's leg bandage. The thick paste won't soak through to irritate the leg, and your horse won't chew it because of the disagreeable taste. When you wash the bandages later, they'll already be covered in soap and all you'll have to do is thoroughly rinse them.

★ When you wash leg wraps and bandages, do a final rinse in vinegar. The pungent smell and taste will usually keep the horse from chewing on the cloth.

AROUND & ABOUT

Keeping track of the various supplies and information you need when you care for and ride horses can be a challenge at times. Here are some miscellaneous tips that may prove useful in certain situations around the stable and on the road.

★　　★　　★

Tarp Tie-Down Repair

Here's a quick fix for times when you're using a tarp to cover a haystack, horse trailer, or woodpile and some of the metal grommets are pulled out, or you need to tie a spot where there is no grommet. Push a smooth stone, a small ball, or a wad of baling twine into the tarp from the other side. Tie a rope or twine tightly around the base of the bulge, tarp and all, to create a fixed, solid location to which you can secure a tie rope. Wrap the rope or twine around a couple of times and then tie a good knot that will stay tight whether or not it has tension on it (see chapter 9 for information on various types of knots).

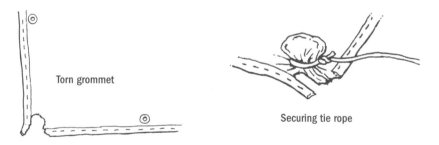

Torn grommet

Securing tie rope

Easy Tool-Handle Repair

You never know when you might break or crack the handle of your pitchfork, manure fork, rake, or shovel, perhaps by accidentally dropping a hay bale from the stack onto it or running over it with a vehicle. Wrap the damaged handle with duct tape or, for an even better repair, use rubber grip tape made for tennis racket handles (and usually sold at sporting goods stores for less than the price of a new tool handle). You can use the grip tape by itself or on top of duct tape.

SOFT BUCKET HANDLE

If you regularly carry feed or water buckets to your horse, putting a fleece halter cover over the handle will make hauling easier on your hands, particularly when the bucket is heavy.

Duct tape or grip tape also improves wood handles on older tools and wheelbarrows that have become rough and weathered. It'll make the surface smoother so you won't get splinters in your hands.

Winter Chore Mittens

Although mittens are warmer than gloves, they hinder the use of your fingers. To keep your hands warm in winter and still have fingers free when you need to buckle a halter or do any other intricate task, use a pair of hunting mittens. They have a slit in the palm to get your fingers out. As an alternative, cut a similar slit in a pair of work mittens; you can keep your fingers warm except for the brief times you need to have them free.

Safer Night Riding

If you ride late in the day and know you may be coming home along a road after dark, a good way to make sure you and your horse are visible to motorists is to use reflector bands made for bicyclists. The silver and orange straps have Velcro fasteners that are designed to go around human arms or legs but fit nicely around a horse's pastern. Using them on all four legs will enable drivers approaching from the front or the rear to gauge how much room to give you. You can also put one around the noseband of a bridle, or on the halter under the bridle.

Drying/Storage Boot Rack

You can create a handy boot rack from a 6-inch-wide board by cutting out appropriately sized circular sections to hang boots in, soles up. Attach the rack to a wall with brackets, so the board is perpendicular to the wall and the boots don't touch the floor. Hung straight in this fashion, the leathers of tall boots won't bend or fold.

Homemade Rope and Handle Protectors

For an inexpensive, safe stall guard to place across a stall opening, fit a length of worn-out garden hose over a small-diameter nylon rope, such as ski tow rope. Tie snaps to the ends of the rope, and attach them to eyebolts at the stall opening. Use several of these hose-covered barriers at appropriate heights to keep the horse in the stall.

An old hose also makes good protective encasement for bucket handles or metal edges on a trailer manger. Warm the hose and slit it, then force it over the handle or edge. As it cools, it will contract and stay in place.

Measuring a Horse's Height

A measuring stick for horses consists of a tall stick marked with 4-inch increments (hands) and a crosspiece that goes over the withers and can be adjusted up or down. If you don't have one of these, you can use a carpenter's tape to measure from the ground up to the height of a straight stick held over the highest point of the withers when the horse is standing squarely on a flat surface. Alternatively, you can drop a string from

---

ARENA MARKERS

To make dressage markers or any other kind of spot markers for a riding ring, rinse out plastic gallon milk jugs and fill them about halfway with sand or fine gravel to make them heavy enough not to blow around or tip over. Write numbers or letters on the sides with a black marking pen.

DRIP-DRY SPONGES

A convenient way to let sponges dry after washing your horse or tack is to put them in a mesh bag (the kind onions come in from the grocery store) and hang it up in a sunny area.

the straight stick (making sure you're holding it level), mark the string where it hits the ground, and then measure the string.

Once you have the height in inches, divide by four to see how tall your horse is in hands. For instance, a horse 60 inches tall is 15 hands, and a horse 63 inches tall is 15.3 hands.

Measuring Bone

A horse with "good bone" (sturdy leg bones) is more likely to stay sound when used hard than a horse with light or fine bone. The amount of bone a horse has, in horseman's terms, is really a combination of the cannon bone and the tendons behind it, and it's determined by measuring around the leg just below the knee. A horse with tendons set well back from the cannon bone has a greater total circumference of limb than a horse with tendons tied in too close, and there is less friction between the moving parts, and they hold up better.

The average domestic horse needs about 8 inches of bone per 1,000 pounds of body-weight. That means a 1,200-pound horse should have at least 9½ inches of bone to provide good support for his body. Domestic horses today commonly have too little bone; we have developed horses that are larger than their wild ancestors but without much increase in bone size to accommodate the additional weight. When choosing a horse, remember this factor.

Creating a Paper Trail

It's frustrating to arrive at a horse show or some other event and discover you've left your horse's registration papers, health certificate, or other necessary papers at home. To avoid making this mistake, keep photocopies of all veterinary records and other pertinent documents in your truck's glove compartment. If you sometimes haul your horse with a friend, make a second set of copies to keep in your tack trunk or something else you always take with you.

IDENTIFICATION NUMBERS

When using a lip tattoo as identification on your horse, or when engraving or marking your tack to prevent theft, use your driver's license number and state abbreviation. That information is much easier for a police or sheriff's department to track down and identify than your name or social security number.

Making Sure Someone Knows Your Plan

When going for a ride in the backcountry, even if it's just for an hour or so off the main roads or trails, travel with a buddy or tell someone where you're going and when you plan to return. You'll be easier to find if you run into any problems on the ride. If you haul your horse to a trailhead, leave emergency contact information on the front seat of your locked vehicle, where someone can find it.

★ ADDITIONAL READING ★

Damerow, Gail. *Fences For Pasture and Garden*. North Adams, MA: Storey Publishing, 1992.

Haas, Jessie. *Safe Horse; Safe Rider*. North Adams, MA: Storey Publishing, 1994.

Hill, Cherry. *Horse Handling and Grooming*. North Adams, MA: Storey Publishing, 1997.

Hill, Cherry. *Horse Health Care*. North Adams, MA: Storey Publishing, 1997.

Hill, Cherry. *Horsekeeping on a Small Acreage* (2nd edition). North Adams, MA: Storey Publishing, 2005.

James, Ruth B., DVM. *How to Be Your Own Veterinarian (Sometimes)*. Mills, WY: Alpine Press, 2000.

Kevil, Mike. *Starting Colts*. Colorado Springs, CO: Western Horseman, 1990.

Mettler, John J. Jr., DVM. *Horse Sense*. North Adams, MA: Storey Publishing, 1989.

Poe, Rhonda Hart. *Trail Riding*. North Adams, MA: Storey, 2005.

Swan, Kathy, and Karan Miller. *Helpful Hints for Horsemen*. Colorado Springs, CO: Western Horseman, 2002.

Thomas, Heather Smith. *Storey's Guide to Raising Horses*. North Adams, MA: Storey Publishing, 2000.

Thomas, Heather Smith. *Storey's Guide to Training Horses*. North Adams, MA: Storey Publishing, 2003.

★ INDEX ★

Numbers in *italics* refer to illustrations; numbers in **boldface** refer to boxes or charts.

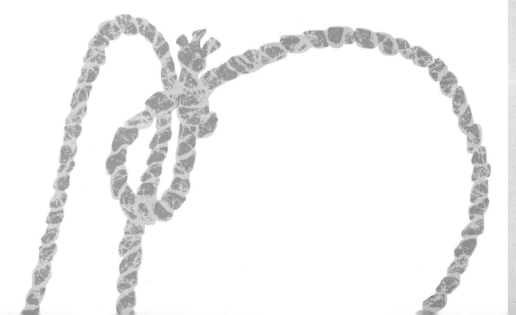

★ OTHER STOREY TITLES YOU WILL ENJOY ★

The Horse Conformation Handbook, by Heather Smith Thomas. Horse conformation —how the shape and form of a particular horse's body compares to an anatomical ideal — affects a horse's performance and suitability for particular functions. This detailed "tour of the horse" analyzes all aspects of conformation that are critical for every horse owner to understand. 400 pages. Paperback. ISBN 1-58017-558-9.

Storey's Guide to Raising Horses, by Heather Smith Thomas. Whether you are an experienced horse handler or are planning to own your first horse, this complete guide to intelligent horsekeeping covers all aspects of keeping a horse fit and healthy in body and spirit. 512 pages. Paperback. ISBN 1-58017-127-3.

Storey's Guide to Training Horses, by Heather Smith Thomas. This comprehensive guide covers every aspect of the training process-from basic safety to retraining a horse that has developed a bad habit-this is an essential handbook for all horse owners. 512 pages. Paperback. ISBN 1-58017-467-1.

Storey's Guide to Feeding Horses, by Melyni Worth. This comprehensive handbook provides down-to-earth explanations of everything from how a horse digests its food to why the horse's summer diet is not suitable when winter comes roaring in. 240 pages. Paperback. ISBN 1-58017-492-2.

The Horse Behavior Problem Solver, by Jessica Jahiel. Using a friendly question-and-answer format and drawing on real-life case studies, Jahiel explains how a horse thinks and learns, why it acts the way it does, and how you should respond. 352 pages. Paperback. ISBN 1-58017-524-4.

Horsekeeping on a Small Acreage, by Cherry Hill. Thoroughly updated, full-color edition of the best-selling classic details the essentials for designing safe and functional facilities whether on one acre or one hundred. Hill describes the entire process: layout design, barn construction, feed storage, fencing, equipment selection, and much more. 320 pages. Paperback. ISBN 1-58017-535-X.

These books and other Storey books are available whereever books are sold, or directly from Storey Publishing, 210 Mass MoCA Way, North Adams, MA 01247, or by calling 1-800-441-5700. www.storey.com